Dynamical Symmetries and Chaotic Behaviour in Physical Systems

Proceedings of the ENEA Workshops on Nonlinear Dynamics, Volume 1

Dynamical Symmetries and Chaotic Behaviour in Physical Systems

ENEA, Bologna, Italy 20-21 June 1989

Editors

Giuseppe Maino
ENEA, Bologna, Italy

Leone Fronzoni
Physics Department
University of Calabria
Cosenza, Italy

Marco Pettini
Osservatorio Astrofisico di Arcetri
and INFN, Firenze, Italy

World Scientific
Singapore • New Jersey • London • Hong Kong

Published by

World Scientific Publishing Co Pte Ltd
5 Toh Tuck Link, Singapore 596224
USA office: 27 Warren Street, Suite 401-402, Hackensack, NJ 07601
UK office: 57 Shelton Street, Covent Garden, London WC2H 9HE

British Library Cataloguing-in-Publication Data
A catalogue record for this book is available from the British Library.

DYNAMICAL SYMMETRIES AND CHAOTIC BEHAVIOUR IN PHYSICAL SYSTEMS
Proceedings of the ENEA Workshops on Nonlinear Dynamics, Volume I

ISBN-13 9789810202965
ISBN-10 9810202962

Preface

This volume is divided into two main parts which contain the texts of invited lectures given at the first two workshops on mathematical physics and non-linear dynamics organized by ENEA (Italian Commission for Nuclear and Alternative Energy Sources), and held at Bologna (Italy) on June, 20/21 and October, 23/24, respectively.

The total attendence of 99 scientists from Universities and National Research Laboratories certainly attests the high interest in this area and the broad range of applications in many branches of pure and applied physics. The scientific program included 18 overview talks, 13 of them are here collected.

The first part deals with general presentations about dynamical (super)symmetries and spectrum-generating algebras. Group-theory techniques are now widely used for the description of a variety of physical phenomena ranging from condensed matter to classical and quantum optics, to particle, atomic and nuclear physics. Surveys are given of the current status of research in these fields with emphasis on selected topics of particular significance.

The second half of the book focuses reader's attention on aspects directly related to non-linear dynamics and chaotic behaviours of physical systems. Relevant contributions resolve to give a comprehensive view to subjects in classical fields of optics and electromagnetism, with regards to important problems like beam stability in particle accelerators, laser oscillators, optical waveguides, etc.. The following two volumes in this series will cover other remarkable and updated applications of non-linear and stochastic dynamics to the physics of fluids and plasmas, and simulation techniques for complex systems.

The effective interest of prof. G.Stoppini (University of Pisa and ENEA) and dr. F.Pistella (ENEA general manager) is gratefully acknowledged. Several discussions with prof. A.Mathis (ENEA Rome, Area Energia e Innovazione) gave rise to this initiative. In particular, we would like to thank dr. E.Cupini, head of ENEA Computing Division, for his kind encouragement in the organization of these workshops. Thanks are also due to prof. M.Rasetti (Politecnico di Torino) for his invaluable advice.

The conference organization benefitted by the support given by drs. O.Bartoli, A.Luise and T.Volpe of ENEA 'Unita' di Progetto Formazione', whose assistance throughout the workshops is appreciated. On-site the conference was aided by Mrs. M.Mazzacurati who was involved in both planning and actual running of it.

The Editors

CONTENTS

FOREWORD

Most natural systems and phenomena are described by non linear equations.
In particular non linear are several differential equations of Classical Dynamics. Often it is possible to obtain solutions in this theory only after an approximated linearization of the equations. In spite of these approximations linear analysis applied to Dynamics has led to spectacular successes in Science and Technology.
Today we are faced with complex systems whose modelization and simulation seems to be impossible by using a linear analysis. It is well known that in physics fluids and plasmas require a non linear approach. Recent observations have furnished further evidence of non linear phenomena in optics and condensed matter physics. Several facts collected in the last twenty years have led physicists and mathematicians to develop a new analytic framework now called Non Linear Dynamics. The new methods have been used, often with success, in Sciences different from physics ranging from chemistry to ecology, from thermodynamics to economics, from neurophysiology to social sciences.
Linear dynamical systems typically show a motion that can be described in terms of well behaved functions that undergo small variations for small changes of the parameters of the equations.
Non linear systems, however, often show transitions from smooth motion to chaotic, erratic or random behaviour. These qualitative changes of the solutions can result from small variations of the parameters. It is the most fascinating look of non linear dynamics.
The first approach to non linear dynamics is due to Poincarè, at the turn of this century. He found that changing some parameter the unique solution of

a non linear equation could split in more solutions. In other words for a given value of the parameter a stable solution could pass to an oscillation. The transition points are called <u>bifurcation</u> points.
After Poincarè and around 1930, a Russian group (Lyapunov, Andronov) was able to develop a theory of non linear oscillations that enjoyed of several interdisciplinary applications.
In 1958 Belusov successfully used non linear methods to explain the oscillations observed in a particular chemical process and in the same period Krylov and Bogoliubov were able to explain phenomena of non linear optics observed in lasers.
In 1963 studying the instabilities of atmosphere E. N. Lorenz numerically found nonperiodic (erratic) solutions of a system of three non linear ordinary differential equations; moreover he introduced a first <u>return map</u>, a discrete transformation by which many qualitative features of the solutions of the system could be reproduced.
A typical map of the interval, like the logistic one, shows the existence of stable points where all solutions are attracted no matter what the initial values are (attractors).
For given values of the parameter (bifurcations) the solutions split showing the existence of attracting limit cycle. At each bifurcation the period of the cycle doubles. Above a critical value of the parameter for many (but not all) of its values the attracting set shows no periodicity whatsoever. In other words at the critical value there is a transition to chaos.
All these properties were established by Ulam, von Neumann, Kac, Metropolis, Stein.
In 1977 Feigenbaum discovered the group properties of the attracting cycles and the existence of scaling laws.

From the Red Spot of Jupiter through clumps of electromagnetic radiation in turbulent plasmas to microscopic charge density waves on solids, spatially localized, long-lived wave-like excitations abound in non linear physical systems. In all these cases coherent structures arise, that represent the natural modes for understanding the time-evolution of a non-linear system and often dominate the long-time behavior of the motion.
Solitons are the most remarkable example of these structures.
Their theory, due to Kruskal and Zabusky (1965), led to the explanation of the solitary waves observed by a satellite in the Adaman Sea and to describe the observed charge waves in solids. In fiber optics, conducting polymers, Josephson transmission lines and plasma cavitons, the prevailing mathematical models are slight modifications of soliton equations.
In an optical fiber transmitting high power, the propagation becomes non linear and solitary wave pulses are formed.
An optical system with Raman amplifiers can use solitons of 20 picoseconds duration to send information at a bit rate above 10 GHz, ten times the conventional rate. The method, tested in laboratory, represents an exciting and real possibility for future long-distance communication.
More general coherent structures can have an essential role in dynamical systems. Among them we mention vortices and fronts.
Well known examples of fluid vortices are the giant Red Spot of Jupiter, tornados in Earth's atmosphere, modons in the Gulf stream current, rotons in superfluid helium, the vortex pattern formed in the atmosphere by a moving body.
The existence of fronts as coherent structures provides another illustration of the important role of non linearity in the technological world. For example, chemical reaction-diffusion systems can

have traveling wave fronts separating reacted and unreacted species. These traveling chemical waves are coupled with fluid modes as in flame fronts or in internal combustion engines. Concentration fronts arise in the leaching of minerals from ore beds while in advanced oil recovery processes unstable fronts between the injected water and the oil trapped in the reservoir control the effectiveness of the recovery process.
The list could be furtherly enriched and it appears that an extensive use of the non linear mathematics can help not only to improve the insight on basic processes but also to stimulate further and important technological advancements.

In the past twenty years several non linear phenomena have been found in biology, ecology, dynamics of population, propagation of epidemics etc.
Many used the logistic map to study the growth of population, moving fronts between infected and non-infected individuals were identified and described in the epidemiology of deseases such as rabies.
The use of a non linear approach has led Prigogine and the Bruxelles school to a "different" thermodynamics.
It seems that all these observations can be explained and described by using a "universal" non linear language so to support the belief that non linear mathematics can be considered the fundament of a new science, the Non Linear Science able to solve or clarify most of the problems inherent to the wide class of complex systems described by non linear equations and whose solution, then, cannot be built as a linear superposition of the solutions inherent to all the subsystems.
In 1975 a group of researchers proposed for non linear Science the name of Synergetics to underline its unifying potential toward several different disciplines.

There are two reasons to explain why this new line only recently began to grow.
The first and most crucial one is bound to the effect of the high-speed computers developed only recently. They allow a quantitative numerical simulation of non linear systems and offer the possibility of discovering new properties more than confirming ideas anticipated by a qualitative analysis.
The second reason has been the observation of universal non linear characteristics in natural systems. In the past two decades the experiments have reached a precision such to allow the observation of quantitative similarities among several non linear processes such as, for example, the route to chaos among an enormous variety of non linear systems.
Why an interest of ENEA, the Italian Commission for Nuclear and Alternative Energy Sources, toward this newly growing matter?
The first reason is that ENEA is increasingly involved into the study, modeling and simulation of systems with growing complexity. These problems range from plasma at the ignition, through the analysis and diagnosis of an environment and the simulation of advanced materials to the behavior of biological systems.
The second reason is connected with the easiness for ENEA to stimulate interdisciplinary activities added to the availability of high computer facilities.
The third reason is represented by the qualified presence of ENEA researchers in the development of non linear dynamics as it is largely demonstrated by the number and quality of their contributions to this volume and the forthcoming ones.
This book collects a series of lectures conceived and planned by dr. E. Cupini and held in ENEA Bologna, Area Energia e Innovazione. Aim of the

lectures was to establish the quality and dimension of the Italian competence on non linear dynamics, to stimulate collaborations, to help to single out a possible ENEA program of activities in the field.

Unvaluable has been the contribution of several Professors and Researchers of the University. Avoiding to mention them individually, I thank all for their enthusiastic partecipation, with the hope that their collaboration with ENEA will be extended in the future. Thanks are also due to Dr. A. Mathis ENEA Roma Area Energia e Innovazione, for several discussions which have been at the origin of this and other initiatives of ENEA on non-linear dynamics.

GHERARDO STOPPINI
(Professor of general physics at the University of Pisa and member of the ENEA Administrative Council).

PART I: DYNAMICAL SYMMETRIES AND APPLICATIONS.

THE CLUSTER FERMI-LINEARIZED HUBBARD MODEL

Arianna Montorsi[♮] and *Mario Rasetti*[♮]

Abstract

The spectrum of the Hubbard model hamiltonian is constructed, in the frame of a self-consistent fermionic linearization scheme, for two-site (dimer) and three-site (trimer) clusters. It is shown that the resulting model has a dynamical algebra which is a superalgebra, whose structure is briefly discussed, and that thermal averages can be reconducted to expectations in the superalgebra generalized coherent states.

♮ *Dipartimento di Fisica and Unitá I.N.F.M., Politecnico di Torino, 10100 Torino, Italy*

The interest in the 2-d Hubbard model [1] has recently been revived by the crucial role it plays in the description of the electrons in the Copper-Oxide planes of a class of high-T_c superconductor ceramic materials [2],[3].

In its customary form the hamiltonian for the model is written

$$H_{Hub} = \sum_{\mathbf{i}\in\Lambda,\sigma} \epsilon_{\mathbf{i}} n_{\mathbf{i},\sigma} + \sum_{\mathbf{i}\in\Lambda} U_{\mathbf{i}} n_{\mathbf{i},\Uparrow} n_{\mathbf{i},\Downarrow} + \sum_{<\mathbf{i},\mathbf{j}>} t_{\mathbf{i},\mathbf{j}} \sum_{\sigma} a^{\dagger}_{\mathbf{i},\sigma} a_{\mathbf{j},\sigma} \quad ; \tag{1}$$

in terms of fermionic operators

$$\{a_{\mathbf{i},\sigma}, a_{\mathbf{j},\sigma'}\} = 0 \quad ; \quad \{a_{\mathbf{i},\sigma}, a^{\dagger}_{\mathbf{j},\sigma'}\} = \delta_{\mathbf{i},\mathbf{j}}\delta_{\sigma,\sigma'} \quad ; \quad n_{\mathbf{i},\sigma} := a^{\dagger}_{\mathbf{i},\sigma} a_{\mathbf{i},\sigma} \quad , \tag{2}$$

where $\mathbf{i} \in \Lambda$ is the site label (Λ denoting the d-dimensional lattice) whereas $\sigma \in \{\Uparrow, \Downarrow\}$ is the spin. $\varepsilon_{\mathbf{i}}$ and $U_{\mathbf{i}}$ represent, respectively, self-energy minus chemical potential and Coulomb repulsion at site $\mathbf{i}$, while $t_{\mathbf{i},\mathbf{j}} = t^*_{\mathbf{j},\mathbf{i}}$ is the hopping amplitude between sites $\mathbf{i}$ and $\mathbf{j}$. The sum over $< \mathbf{i},\mathbf{j} >$ is meant to run over all sites $\mathbf{i}$ and $\mathbf{j}$ of Λ which are nearest neighbors ($n.n.$).

No exact solution is known for the model described by (1) for $d = 2$, but an increasing amount of numerical as well as of approximate results [4],[5] have provided interesting information, particularly in the limit $|U/t| \to \infty$ and in proximity of band half-filling (*i.e.* average on-site occupation number equal to 1). In a set of recent papers [6]÷[9] a different approach was proposed, which resorts to fermionic linearization. The latter was extended to two-site clusters, within a scheme which allows a deeper insight in the behaviour of the order parameters, leads to the explicit construction of the phase diagram, and – due to its algebraic structure [9] – lends itself to a solution which might possibly be reconducted, by a limiting procedure, to be the exact one for the original model. In the scheme proposed, the spectrum is obtained – after recognizing that the linearized model has a dynamical algebra which is a $\mathbb{Z}_2$-graded algebra (*superalgebra*) – via an inner automorphism of the algebra itself (which generalizes the customary Bogolubov rotation). The self-consistency equations reconduct the exact results for the linearized model to approximate (mean-field like) results for the model (1). The solution of such equations moreover allows computing statistical mechanical averages, from which, in turn, one can straightforwardly derive the structure of phase space.

In the present note we derive within a self-consistent scheme the spectrum of the hamiltonian for both the two-site (*dimer*) and the three-site (*trimer*) clusters. We focus our attention on the spectrum, in that this is the key object from which all relevant properties, such as the ground state , the zero-temperature phase diagram and the physically relevant order parameters, can be derived (see ref. [8]). It is worth pointing out that the trimer case we consider also includes in the hamiltonian a next-nearest neighbour hopping term.

The cluster-linearization method consists in taking into account exactly all the two-body interactions within the selected cluster, and using the fermi linearization to describe all hopping interactions between neighbouring clusters. This implies that the latter hopping terms [7],[8] $a^{\dagger}_{i,\sigma}a_{j,\sigma}$ of the hamiltonian (1), bilinear in the fermion operators $a^{\dagger}_{i,\sigma}$ and $a_{j,\sigma}$, for which i and j denote sites belonging to *n.n.* clusters, are replaced by the linear form

$$a^{\dagger}_{i,\sigma}a_{j,\sigma} \mapsto \bar{\eta}_{i,\sigma}a_{j,\sigma} + a^{\dagger}_{i,\sigma}\eta_{j,\sigma} - \bar{\eta}_{i,\sigma}\eta_{j,\sigma} \quad , \tag{3}$$

where the coefficients $\eta_{j,\sigma}, \bar{\eta}_{i,\sigma}$ are variables anticommuting with both the fermion operators and among themselves ($\{\eta_i, \eta_j\} = 0$) [the indexes i, j stand here for the pairs (i,σ) and (j,σ) respectively]. They belong to the *odd* sector $\mathcal{G}_o$ of a *non-associative* $\mathbb{Z}_2$-graded algebra $\mathcal{G} = \mathcal{G}_o \oplus \mathcal{G}_e$, defined by $[\mathcal{G}_o, \mathcal{G}_e] = 0$; $\mathcal{G}_e$ commuting, and $\mathcal{G}_o$ anticommuting with products of odd number of fermion operators, both commuting with products of even number of fermion operators; and $\mathcal{G}_e$ containing all products of even number of elements of $\mathcal{G}_o$ ($\mathbb{R} \subseteq \mathcal{G}_e$). Non-associativity raises from the physical requirement that $\eta_i^2 = 0$ but $(\eta_i\eta_j)^2 \neq 0$. For the sake of simplicity, we shall also assume that $\eta_j \equiv \eta_j$, independent on σ.

A realization of the elements of $\mathcal{G}_e$, adopted in ref. [8], consists in the identification of the product in $\mathcal{G}$ with the following inner product over the vector (*super-*)space associated with $\mathcal{G}$:

$$\eta_i\eta_j \equiv \sin(\varphi_i - \varphi_j) \quad , \tag{4}$$

where we denote by φ_i the phase associated with η_i (of course $-\varphi_i$ will be that of $\bar{\eta}_i$), which has to be determined consistently with the relations $\eta_i\bar{\eta}_i \equiv \sin(2\varphi_i)$.

The total hamiltonian in the cluster linearization scheme reads then $H =$

$\sum_{\{c\}\in\Lambda} H_c$, where $\{c\}$ denotes any possible cluster covering of Λ, and

$$H_c = \sum_{\alpha=1}^{\nu_c} H_\alpha^\ell + \kappa_c \sum_{\substack{\alpha=1 \\ \mathrm{mod}\,\nu_c}}^{\nu_c} (A_\alpha^\dagger A_{\alpha+1} + A_{\alpha+1}^\dagger A_\alpha) \quad ; \tag{5.1}$$

where $\kappa_c \equiv (\nu_c - 1)\,t$, and

$$A_\alpha = \frac{1}{\sqrt{2}}(a_{\alpha,\Uparrow} + a_{\alpha,\Downarrow}) \quad . \tag{5.2}$$

$\nu_c\,(=2,3)$ denotes the number of sites in cluster c, and for simplicity we assumed the *hopping amplitude* to be real and to have the same value t for all pairs of *n.n.* sites[‡]. As we are interested here only in the spectrum of the hamiltonian and in the zero temperature features of the model , it is irrelevant which covering one selects (whereas at $T \neq 0$ one should consider all possible dimer coverings of the lattice in order to account properly for the global lattice entropy). It is worth to point out that the trimer is the minimal cluster necessary to obtain a complete covering of the decorated lattice typical of Copper-Oxide planes in high-T_c superconductors.

In eq. (5.1),

$$H_\alpha^\ell = \varepsilon_\alpha(n_{\alpha,\Uparrow} + n_{\alpha,\Downarrow}) + U_\alpha n_{\alpha,\Uparrow} n_{\alpha,\Downarrow} + \sqrt{2}(\vartheta_\alpha A_\alpha + A_\alpha^\dagger \bar{\vartheta}_\alpha) - C_\alpha \tag{6}$$

with

$$C_\alpha \doteq \frac{1}{2}\kappa_c \Big\{ \sum_{\mathrm{i}\in\{\mathrm{i}\in\Lambda\backslash c\,|\,\mathrm{i}=\mathrm{n.n.}(\alpha)\}} (\bar{\eta}_\alpha \eta_\mathrm{i} + \bar{\eta}_\mathrm{i} \eta_\alpha) - 2 \sum_{\substack{\gamma=1 \\ \gamma\neq\alpha}}^{\nu_c} \bar{\eta}_\alpha \eta_\gamma \Big\} \mathbf{I} \quad , \tag{7.1}$$

and

$$\vartheta_\alpha := \sum_{\mathrm{i}\in\{\mathrm{i}\in\Lambda\backslash c\,|\,\mathrm{i}=\mathrm{n.n.}(\alpha)\}} t_{\mathrm{i},\alpha} \bar{\eta}_\mathrm{i} - \sum_{\substack{\gamma=1 \\ \gamma\neq\alpha}}^{\nu_c} t_{\gamma,\alpha} \bar{\eta}_\gamma \quad . \tag{7.2}$$

For the sake of simplicity we have also included spin-flip hopping processes between sites in the cluster c with the same amplitude as parallel spin hoppings[♮].

---

‡ Notice that in the trimer case one of the pairs which are nearest neighbours on the cluster are indeed next-nearest neighbours on the lattice.

♮ Notice that the possible diversity in the two amplitudes has been embodied into the normalization of C_α in (7.1) and of the ϑ_α's in (7.2).

H_c as given by (5.1),(6) can be straightforwardly recognized as an element of the superalgebra $\mathcal{S}_c = \oplus_{\lambda\in\mathcal{L}} u(\nu_c|1)_\lambda$. The index λ assumes $|\mathcal{L}| = 2^{\nu_c}$ different values, where we denote by $\mathcal{L}$ the set of all possible λ's. Each $u(\nu_c|1)_\lambda$ equals $\mathcal{P}_\lambda u(\nu_c|1)$, where the $\mathcal{P}_\lambda$'s are the projection operators defined below, whereas $u(\nu_c|1)$ is a well known superalgebra [10] generated by $(\nu_c+1)^2$ elements, (ν_c^2+1) of which form the bosonic sector $\mathcal{B}(u(\nu_c|1))$ and $2\nu_c$ are in the fermionic sector $\mathcal{F}(u(\nu_c|1))$:

$$\begin{aligned} \mathcal{B}(u(\nu_c|1)) &= \{\mathbf{I}, A^\dagger_\alpha A_\gamma \,;\, \alpha,\gamma = 1,\ldots,\nu_c\} \quad ; \\ \mathcal{F}(u(\nu_c|1)) &= \{A_\alpha, A^\dagger_\alpha \,;\, \alpha = 1,\ldots,\nu_c\} \quad . \end{aligned} \tag{8}$$

The 2^{ν_c} orthogonal copies of $u(\nu_c|1)$ which generate the whole $\mathcal{S}_c$ can be obtained by multiplying the odd and even sectors of $u(\nu_c|1)$, as given in (8), successively by the 2^{ν_c} inequivalent idempotent elements in the center C of $\mathcal{S}_c$:

$$\mathcal{P}_\lambda \equiv \prod_{\alpha=1}^{\nu_c} p_\alpha^{(\lambda)} \quad ; \tag{9.1}$$

where, writing the multi-index λ as the collection of binary digits $\{d_\alpha \,|\, d_\alpha \in \{0,1\}\,;\, \alpha = 1,\ldots,\nu_c\}$, $p_\alpha^{(\lambda)}$ reads

$$p_\alpha^{(\lambda)} = \frac{1}{2} e^{-id_\alpha \frac{\pi}{2}} \left(e^{i\frac{\pi}{2}D_\alpha} + e^{id_\alpha\pi} e^{-i\frac{\pi}{2}D_\alpha}\right) \quad ; \tag{9.2}$$

i.e.

$$\begin{aligned} &\text{if } d_\alpha = 0 \quad p_\alpha^{(\lambda)} = \mathbf{I} - D_\alpha \,; \\ &\text{if } d_\alpha = 1 \quad p_\alpha^{(\lambda)} = D_\alpha \,; \end{aligned}$$

where

$$D_\alpha := \frac{1}{2}(n_{\alpha,\Uparrow} + n_{\alpha,\Downarrow} + a_{\alpha,\Downarrow}a^\dagger_{\alpha,\Uparrow} + a_{\alpha,\Uparrow}a^\dagger_{\alpha,\Downarrow}) \quad ; \tag{10}$$

As D_α's are projectors ($D_\alpha^2 = D_\alpha$), the 2^{ν_c} copies of $u(\nu_c|1)$ which generate $\mathcal{S}_c$ are in fact labelled by the 2^{ν_c} possible different combinations of eigenvalues 1 and 0 of $D_\alpha\,;\, \alpha = 1,\ldots,\nu_c$.

The hamiltonian H_c as given in (5.1), (6) can be conveniently written, in terms of the operators D_α and $N_\alpha \equiv A^\dagger_\alpha A_\alpha$. In order to do this, we write first

$$H_c = \sum_{\lambda\in\mathcal{L}} H_\lambda \quad ; \tag{11}$$

where

$$H_\lambda \equiv H_c\,\mathcal{P}_\lambda = \left(H^{(\mathcal{B})} + H^{(\mathcal{F})} + C\right)\mathcal{P}_\lambda \equiv H_\lambda^{(\mathcal{B})} + H_\lambda^{(\mathcal{F})} + C_\lambda \quad . \tag{12}$$

Then, in (12),

$$H^{(\mathcal{B})} = \sum_{\alpha=1}^{\nu_c} \left[\varepsilon_\alpha (D_\alpha + N_\alpha) + U_\alpha D_\alpha N_\alpha\right] + \kappa_c \sum_{\substack{\alpha=1 \\ \mathrm{mod}\, \nu_c}}^{\nu_c} (A_\alpha^\dagger A_{\alpha+1} + A_{\alpha+1}^\dagger A_\alpha) \quad ; \tag{13}$$

$$H^{(\mathcal{F})} = \sqrt{2} \sum_{\alpha=1}^{\nu_c} \left(\vartheta_\alpha A_\alpha + A_\alpha^\dagger \bar{\vartheta}_\alpha\right) \quad , \tag{14}$$

and $C = \sum_{\alpha=1}^{\nu_c} C_\alpha$.

In view of the existence of a dynamical algebra, the spectrum of H_c can be straightforwardly obtained by a generalized Bogolubov rotation [7],[11] in the (super)group space, which in turn can be implemented by the adjoint action of a suitable operator Z in $\mathcal{S}_c$.

With no loss of generality we write

$$\exp Z = \exp Z^{(b)} \exp Z^{(f)} \quad ; \tag{15}$$

where, due to the direct sum structure of $\mathcal{S}_c$, we can set

$$\begin{aligned} Z^{(b)} &= \sum_{\lambda \in \mathcal{L}} \mathcal{P}_\lambda \sum_{\substack{\alpha=1 \\ \mathrm{mod}\nu_c}}^{\nu_c} z_\alpha^{(\lambda)} (A_\alpha^\dagger A_{\alpha+1} - A_{\alpha+1}^\dagger A_\alpha) = \sum_{\lambda \in \mathcal{L}} Z_\lambda^{(b)} \quad ; \\ Z^{(f)} &= \sum_{\lambda \in \mathcal{L}} \mathcal{P}_\lambda \sum_{\alpha=1}^{\nu_c} (\zeta_\alpha^{(\lambda)} A_\alpha + \bar{\zeta}_\alpha^{(\lambda)} A_\alpha^\dagger) = \sum_{\lambda \in \mathcal{L}} Z_\lambda^{(f)} \quad ; \end{aligned} \tag{16}$$

with $\{z_\alpha^{(\lambda)} | \lambda \in \mathcal{L};\, \alpha = 1, \dots, \nu_c\} \in \mathcal{G}_e$, $\{\zeta_\alpha^{(\lambda)} | \lambda \in \mathcal{L};\, \alpha = 1, \dots, \nu_c\} \in \mathcal{G}_o$.

The diagonal form $\tilde{\mathcal{H}}_c$ of H_c can then be obtained from

$$\mathcal{H}_c = \bigoplus_{\lambda \in \mathcal{L}} \exp\left(\mathrm{ad} Z_\lambda^{(b)}\right) \left(\exp\left(\mathrm{ad} Z_\lambda^{(f)}\right)(H_\lambda)\right) = \bigoplus_{\lambda \in \mathcal{L}} \exp\left(\mathrm{ad} Z_\lambda^{(b)}\right) \left(\mathcal{H}_\lambda^{(f)}\right) \quad ; \tag{17}$$

by suitably chosing the coefficients $z_\alpha^{(\lambda)}$ and $\zeta_\alpha^{(\lambda)}$ in such a way that the resulting operator belongs to the Cartan subalgebra of $\mathcal{B}(\mathcal{S}_c)$.

A straightforward calculation gives

$$\mathcal{H}_\lambda^{(f)} = H_\lambda^{(\mathcal{B})} + \tilde{H}_\lambda^{(\mathcal{F})} + \tilde{C}_\lambda \quad , \tag{18}$$

with

$$\tilde{H}_\lambda^{(\mathcal{F})} = \sum_{\substack{\alpha=1 \\ \mathrm{mod}\,\nu_c}}^{\nu_c} \left\{ \left[\vartheta_\alpha + E_\alpha^{(\lambda)} \zeta_\alpha^{(\lambda)} + \kappa_c \left(\zeta_{\alpha-1}^{(\lambda)} + \zeta_{\alpha+1}^{(\lambda)} \right) \right] A_\alpha \mathcal{P}_\lambda + \mathrm{h.c.} \right\} \quad , \tag{19}$$

where we denoted $E_\alpha^{(\lambda)} \equiv \varepsilon_\alpha + U_\alpha d_\alpha$, and

$$\tilde{C}_\lambda = C_\lambda + \sum_{\substack{\alpha=1 \\ \mathrm{mod}\,\nu_c}}^{\nu_c} \left\{ \left[E_\alpha^{(\lambda)} \zeta_\alpha^{(\lambda)} + \kappa_c \left(\zeta_{\alpha-1}^{(\lambda)} + \zeta_{\alpha+1}^{(\lambda)} \right) \right] \bar{\zeta}_\alpha^{(\lambda)} - \left(\bar{\vartheta}_\alpha \zeta_\alpha^{(\lambda)} - \vartheta_\alpha \bar{\zeta}_\alpha^{(\lambda)} \right) \right\} \mathbf{I} \quad . \tag{20}$$

It is convenient to select the running variables $\zeta_\alpha^{(\lambda)}$ in (19) in such a way that the hamiltonian $\mathcal{H}_\lambda^{(f)}$ lives in the bosonic sector $\mathcal{B}$ of S_c. The latter condition in fact implies that the chosen $\zeta_\alpha^{(\lambda)}$'s make $\tilde{H}_\lambda^{(\mathcal{F})}$ vanish. This leads to the following system of linear equations (homogeneous over the odd sector $\mathcal{G}_o$ of $\mathcal{G}$ and hence equivalent to a regular linear system over $\mathbb{R}$) :

$$E_\alpha^{(\lambda)} \zeta_\alpha^{(\lambda)} + \kappa_c \left(\zeta_{\alpha-1}^{(\lambda)} + \zeta_{\alpha+1}^{(\lambda)} \right) = -\vartheta_\alpha \quad ; \quad \alpha = 1, \ldots, \nu_c ; \, \mathrm{mod}\, \nu_c \quad . \tag{21}$$

For the sake of simplicity we do not change our notation, however one should keep in mind that from now on whenever we shall write $\zeta_\alpha^{(\lambda)}$ (or its conjugate $\bar{\zeta}_\alpha^{(\lambda)}$) we shall intend the solutions of (21) :

$$\zeta_\alpha^\lambda = - \sum_{\gamma=1}^{\nu_c} \left(M^{-1} \right)_{\alpha,\gamma} \vartheta_\gamma \quad ; \tag{22}$$

where the matrix M has elements

$$M_{\alpha,\gamma} = E_\alpha \delta_{\alpha,\gamma} + \kappa_c (1 - \delta_{\alpha,\gamma}) \in \mathbb{R} \quad . \tag{23}$$

Before we proceed to completing the rotation (17), we notice that as $\tilde{C}_\lambda$ is proportional to the identity $\mathbf{I}$ and $\tilde{H}_\lambda^{(\mathcal{F})}$ is zero, only $H_\lambda^{(\mathcal{B})}$ is affected by the adjoint action of $Z_\lambda^{(b)}$. On the other hand, recalling that the bosonic sector of a superalgebra is a Lie algebra (in present case $\mathcal{B}(S_c) \equiv u(\nu_c) \oplus u(1)$; the $u(1)$ part being due to the property that $\sum_\alpha N_\alpha$ is a constant of motion), it is convenient to introduce the set of auxiliary bond operators $B_\alpha \doteq A_\alpha^\dagger A_{\alpha+1}$ and $B_\alpha^\dagger$, which satisfy the following commutation relations :

$$\begin{aligned} &[B_\alpha, B_\gamma] = \delta_{\alpha,\gamma-1} B_{\gamma+1}^\dagger - \delta_{\alpha,\gamma+1} B_{\gamma-1}^\dagger \quad ; \quad [B_\alpha, B_\gamma^\dagger] = \delta_{\alpha,\gamma}(N_\alpha - N_{\alpha+1}) \quad ; \\ &[N_\alpha, B_\gamma] = (\delta_{\alpha,\gamma} - \delta_{\alpha,\gamma+1}) B_\gamma \quad ; \quad \alpha, \gamma = 1, \ldots, \nu_c \, , \, \mathrm{mod}\, \nu_c \quad . \end{aligned} \tag{24}$$

In terms of these operators the hamiltonian (13) splits into a part which is site dependent and one which is bond dependent, and so does the Bogolubov rotated hamiltonian

$$\begin{aligned}\mathcal{H}_\lambda^{(b)} &\equiv \exp\left(\mathrm{ad}Z_\lambda^{(b)}\right)\left(H_\lambda^{(B)}\right)\\ &= \left\{\sum_{\alpha=1}^{\nu_b}\left(\sum_{n=0}^{\infty}\frac{1}{n!}c_\alpha^{(n)}\right)(B_\alpha + B_\alpha^\dagger) + \sum_{\alpha=1}^{\nu_c}\left(\sum_{n=0}^{\infty}\frac{1}{n!}E_\alpha^{(n)}\right)N_\alpha\right\}\mathcal{P}_\lambda \quad .\end{aligned} \tag{25}$$

where ν_b denotes the number of bonds in the cluster c.

In (25) the coefficients $c_\alpha^{(n)}$ (we don't make the dependence on λ explicit to simplify the notation, and hence we drop it also in the $z_\alpha^{(\lambda)}$) are solutions of the system of ν_b recursion relations (for $n \geq 2$)

$$\begin{aligned}c_\alpha^{(n)} &= 2z_\alpha\left(z_{\alpha+1}c_{\alpha+1}^{(n-2)} - 2z_\alpha c_\alpha^{(n-2)} + z_{\alpha-1}c_{\alpha-1}^{(n-2)}\right)\\ &\quad + z_{\alpha+1}c_{\alpha-1}^{(n-1)} - z_{\alpha-1}c_{\alpha+1}^{(n-1)} \quad ; \quad \forall\alpha = 1,\ldots,\nu_b \quad ;\end{aligned} \tag{26}$$

with initial conditions $c_\alpha^{(0)} = 2t$; $c_\alpha^{(1)} = z_\alpha(E_{\alpha+1} - E_\alpha) + 2t(z_{\alpha+1} - z_{\alpha-1})$. Notice that for the trimer the site-subindex on the r.h.s. of (26) is meant mod ν_b, whereas for the dimer ($\nu_b = 1$) one should take $z_0 = 0 = z_2$. The coefficients $E_\alpha^{(n)}$ are then given by

$$\begin{aligned}E_\alpha^{(n)} &= 2\left(z_\alpha c_\alpha^{(n-1)} - z_{\alpha-1}c_{\alpha-1}^{(n-1)}\right) \quad , \; n \geq 1 \quad ;\\ E_\alpha^{(0)} &= E_\alpha \; ; \; \alpha = 1,\ldots,\nu_c, \bmod \nu_c \quad .\end{aligned} \tag{27}$$

The solution of the recursive equations (26) can be achieved in two successive steps. First one introduces the generating function

$$\mathcal{Q}_\alpha(x) \doteq \sum_{n=0}^{\infty}\frac{c_\alpha^{(n)}}{n!}x^n \quad . \tag{28}$$

It is straightforward to see, on multiplying both sides of (26) by $\dfrac{x^{n-2}}{(n-2)!}$ and performing then $\sum\limits_{n=2}^{\infty}$, that the recursion relations turn into a differential equation for the function $\mathcal{Q}_\alpha$:

$$\begin{aligned}\frac{d^2\mathcal{Q}_\alpha(x)}{dx^2} &= 2z_\alpha z_{\alpha+1}\mathcal{Q}_{\alpha+1}(x) - 4z_\alpha^2\mathcal{Q}_\alpha(x) + 2z_\alpha z_{\alpha-1}\mathcal{Q}_{\alpha-1}(x)\\ &\quad + z_{\alpha+1}\frac{d\mathcal{Q}_{\alpha-1}(x)}{dx} - z_{\alpha-1}\frac{d\mathcal{Q}_{\alpha+1}(x)}{dx} \quad .\end{aligned} \tag{29}$$

The solution of the system (29) is then easily obtained taking the Laplace transform of both sides, whereby the differential equations turn into the following linear system:

$$\begin{aligned}(s^2+4z_\alpha^2)Q_\alpha(s)+(z_{\alpha-1}s-2z_\alpha z_{\alpha+1})Q_{\alpha+1}(s)\\ -(z_{\alpha+1}s+2z_\alpha z_{\alpha-1})Q_{\alpha-1}(s)=s+z_\alpha(E_{\alpha+1}-E_{\alpha-1})\quad,\end{aligned} \tag{30}$$

the unknown $Q_\alpha(s)$ denoting the Laplace transforms of $\mathcal{Q}_\alpha(x)$:

$$Q_\alpha(s)=\int_0^\infty e^{-xs}\mathcal{Q}_\alpha(x)dx\quad, \tag{31}$$

and use has been made of the well known property that the transform of $\dfrac{d^k\mathcal{Q}_\alpha(x)}{dx^k}$ is $s^kQ_\alpha(s)-\sum_{\ell=0}^{k-1}s^{k-\ell-1}\left.\dfrac{d^\ell\mathcal{Q}_\alpha(x)}{dx^\ell}\right|_{x=0}$ and that $\mathcal{Q}_\alpha(0)=c_\alpha^{(0)}$, $\left.\dfrac{d\mathcal{Q}_\alpha(x)}{dx}\right|_{x=0}=c_\alpha^{(1)}$.

System (30) has solutions of the form :

$$Q_\alpha(s)=\left(\mathcal{M}^{-1}\right)_{\alpha,\gamma}\left[s+z_\gamma\left(E_{\gamma+1}-E_\gamma\right)\right]\quad, \tag{32}$$

where the matrix $\mathcal{M}$ has elements

$$\mathcal{M}_{\alpha,\gamma}=\left(s^2+4z_\alpha^2\right)\delta_{\alpha,\gamma}+\left(z_{\gamma-1}s-2z_\gamma z_{\gamma+1}\right)\delta_{\alpha,\gamma+1}-\left(z_{\gamma+1}s+2z_\gamma z_{\gamma-1}\right)\delta_{\alpha,\gamma-1}\,. \tag{33}$$

Notice that the solutions (32) in fact give $Q_\alpha(s)$ as the ratio of two polynomials in s

$$Q_\alpha(s)=\frac{\mathcal{N}_\alpha(s)}{\mathcal{D}(s)}$$

of which the numerator $\mathcal{N}_\alpha(s)$ has degree $2\nu_b-1$ and the denominator $\mathcal{D}(s)$ has degree $2\nu_b$ in s.

Inverse Laplace transform of (32) finally gives for the *dimer* ($\mathcal{D}(s)=s^2+4z^2$):

$$\mathcal{Q}_\alpha(x)=2t\cos\left(2z_\alpha x\right)+\frac{1}{2}\left(E_{\alpha+1}-E_\alpha\right)\sin\left(2z_\alpha x\right)\quad; \tag{34}$$

and for the *trimer* ($\mathcal{D}(s)=s^2(s^2+|z|^2)(s^2+4|z|^2)$, $|z|^2=z_1^2+z_2^2+z_3^2$) :

$$\mathcal{Q}_\alpha(x)=\sum_{n=1}^{4}\frac{\mathcal{N}_\alpha(s_n)}{\mathcal{D}'(s_n)}\,e^{s_nx}+2\,\frac{\mathcal{N}_\alpha'(0)}{\mathcal{D}''(0)}\quad; \tag{35}$$

where s_n are the simple roots $(\pm i|z|\ ,\ \pm i2|z|)$ of the equation $\mathcal{D}(s) = 0$ and the last term at the *r.h.s.* takes into account that $s = 0$ is a double root.

In order to identify $\mathcal{H}_\lambda^{(b)}$ as an element of the Cartan subalgebra of $\mathcal{B}(\mathcal{S}_c)$ one has finally just to select the coefficients z_α in such a way that $\mathcal{Q}_\alpha(1) = 0$; $\alpha = 1, \ldots, \nu_b$. This is straightforwardly done for the dimer, in which case by eq.(34) one obtains ($\alpha = 1$, as $\nu_b = 1$)

$$z_1 = \frac{1}{2}\arctan\left\{\frac{4t}{E_1 - E_2}\right\} \quad , \tag{36}$$

while for the trimer the resulting system is trascendental and has to be solved numerically.

Finally, the spectrum of $H^{(\mathcal{B})}$ is given by :

$$\tilde{\mathcal{H}}_\lambda^{(b)} = \left\{\frac{1}{2}\left[\frac{(E_1 - E_2)^2 - 16t^2}{\sqrt{(E_1 - E_2)^2 + 16t^2}} + (E_2 - E_1)\right](N_2 - N_1) \right. \\ \left. + E_1 N_1 + E_2 N_2\right\}\mathcal{P}_\lambda \quad ; \tag{37.1}$$

in the dimer case, and by :

$$\tilde{\mathcal{H}}_\lambda^{(b)} = \left\{ \sum_{\substack{\alpha=1 \\ \mathrm{mod}\ \nu_c}}^{\nu_c} N_\alpha \left[E_\alpha + \frac{4}{\mathcal{D}''(0)}\left(z_\alpha \mathcal{N}'_\alpha(0) - z_{\alpha-1}\mathcal{N}'_{\alpha-1}(0)\right) \right.\right. \\ \left.\left. + 2\sum_{k=1}^{4}\left(\frac{z_\alpha \mathcal{N}_\alpha(s_k) - z_{\alpha-1}\mathcal{N}_{\alpha-1}(s_k)}{\mathcal{D}'(s_k)}\frac{e^{s_k} - 1}{s_k}\right)\right]\right\}\mathcal{P}_\lambda \quad . \tag{37.2}$$

in the trimer case.

The complete diagonalized hamiltonian of course is the sum of (37) and $\sum_\lambda \tilde{\mathcal{C}}_\lambda \mathcal{P}_\lambda$ with $\tilde{\mathcal{C}}_\lambda$ as given by (20) and (22). The latter term acquires a physical meaning only after the *mean fermionic amplitudes* ϑ_α (which appear bilinearly, namely through terms $\in \mathcal{G}_e$, which are to be treated in our scheme as c-numbers) have been determined self-consistently. In what follows we shall then discuss briefly the self-consistency conditions.

To this effect we notice that self-consistency implies just the identification of η_α with the average $< a_{\alpha,\sigma} >$. As we assumed η_j independent on σ, this amounts to identifying

$$< \vartheta_\alpha A_\alpha > \equiv \sqrt{2}\left\{\sum_{\mathrm{i}\in\{\mathrm{i}\in\Lambda\backslash c\ |\ \mathrm{i=n.n.}(\alpha)\}} t_{\mathrm{i},\alpha}\bar{\eta}_{\mathrm{i}} - \sum_{\substack{\gamma=1 \\ \gamma\neq\alpha}}^{\nu_c} t_{\gamma,\alpha}\bar{\eta}_\gamma\right\}\eta_\alpha \quad . \tag{38}$$

In (38) $< \bullet >$ is the thermal average in the grand-canonical ensemble (obviously $\bullet$ refers to some dynamical variable pertaining to the cluster c, and we only need to define averages using the cluster hamiltonian):

$$< \bullet > \equiv \frac{\mathrm{Tr}\left\{\bullet\, e^{-\beta H_c}\right\}}{\mathrm{Tr}\left\{e^{-\beta H_c}\right\}} \quad . \tag{39}$$

Upon noticing that, due to cyclic invariance of the trace operation, the numerator at the *r.h.s.* of (39) can be written

$$\sum_{\lambda \in \mathcal{L}} e^{-\beta \tilde{\mathcal{C}}_\lambda} \mathrm{tr}_\lambda \left\{ \exp\left(\mathrm{ad} Z_\lambda\right)(\bullet)\, e^{-\beta \tilde{\mathcal{H}}_\lambda^{(b)}} \right\} \quad ,$$

where tr_λ denotes trace over the λ-th copy of $u(\nu_c|1)$; it is straightforward to obtain

$$< \vartheta_\alpha A_\alpha > = - \frac{\sum_{\{d_\alpha\},\{n_\alpha\}} \left\{ \left(\sum_{\gamma=1}^{\nu_c} \vartheta_\alpha (M_{-1})_{\alpha,\gamma} \bar{\vartheta}_\gamma \right) \exp\left[-\beta \left(\tilde{\mathcal{H}}_\lambda^{(b)} + \tilde{\mathcal{C}}_\lambda \right) \right] \right\}}{\sum_{\{d_\alpha\},\{n_\alpha\}} \exp\left\{ -\beta \left(\tilde{\mathcal{H}}_\lambda^{(b)} + \tilde{\mathcal{C}}_\lambda \right) \right\}} \; ; \tag{40}$$

where n_α, and d_α denote the eigenvalues of N_α and D_α, respectively, and use has been made of the property that $\sum_{\lambda \in \mathcal{L}} \mathrm{tr}_\lambda$ equals the trace in the Fock space $\sum_{\{d_\alpha\}} \sum_{\{n_\alpha\}}$. Moreover, in (40) the ϑ_α's should be thought of as functions of the η_α's through (7.2). By comparison of (38) with (40), we finally obtain the self-consistency conditions as a system of equations in the bilinear products $\bar{\eta}_\alpha \eta_\gamma$, which, according to (4), are c-numbers.

We conclude by pointing out that previous scheme is tantamount to representing the thermodynamical states of the system described by the hamiltonian H as generalized supercoherent [8] states for the superalgebra $\mathcal{S}_c$. The latter indeed have the form

$$|z, \zeta > = \exp(Z)|\omega>_c \quad , \quad Z = -Z^\dagger \in \mathcal{S}_c \quad ; \tag{41}$$

where $|\omega>_c$ is the *highest weight vector* of $\mathcal{S}_c$, and Z has the form given in (15) and (16) (obviously with generic z_α's and ζ_α's).

Resorting to such supercoherent states, one can then construct a number of order parameters interesting from the physical point of view, such as e.g.

$< a_{\alpha,\Uparrow} a_{\alpha,\Downarrow} >$, $< a_{\alpha,\sigma} a_{\alpha+1,\sigma'} >$, simply as expectation values in the coherent state ($< z,\zeta| \bullet |z,\zeta >$). We recall that in ref. [8] this was done for the dimer and lead to a quite interesting structure of the phase space.

References

[1] J. Hubbard, *Proc R. Soc. London*, Sec. *A* **276**, 238 (1963) and **277**, 237 (1964)

[2] V.J. Emery, *Phys. Rev. Lett.* **58**, 2794 (1987)

[3] F.C. Zhang, and T.M. Rice, *Phys. Rev.* **B37**, 3759 (1988)

[4] J.E. Hirsch, *Phys. Rev. Lett.* **54**, 1317 (1985) ;
S.R. White, D.J. Scalapino, R.L. Sugar, E.Y. Loh, J.E. Gubernatis, and R.T. Scalettar, *Phys. Rev.* **B40**, 506 (1989)

[5] W. Metzner, and D. Vollhardt, *Phys. Rev.* **B39**, 4462 (1989)

[6] A. Montorsi, M. Rasetti, and A.I. Solomon, *Phys. Rev. Lett.* **59**, 2243 (1987)

[7] A. Montorsi, M. Rasetti, and A.I. Solomon, *Intl. J. Mod. Phys.* **B3**, 247 (1989)

[8] A. Montorsi, and M. Rasetti, *Mod. Phys. Letters* **B**, in press

[9] A. Montorsi, M. Rasetti, and A.I. Solomon, in "Proceedings of the Colloquium on *Group Theoretical Methods in Physics*, Moscow 1990", in press

[10] V.G. Kac *Adv. Math.* **26**, 8 (1977) , and *Commun. Math. Phys.* **53**, 31 (1977)

[11] A.I. Solomon, *J. Phys.***A** : *Math. Gen.* **14**, 2177 (1981)

A general view to Lie algebraic methods in applied mathematics, optics and transport systems for charged beam accelerators.

G. Dattoli and A. Torre

ENEA, Dip. TIB, U.S. Fisica Applicata, CRE Frascati
C.P. 65 - 00044 Frascati (Roma), Italy

Abstract

The theory of partial and ordinary differential equations is reformulated within the context of a unifying formalism, which combines the algebraic ordering procedure with the matrix image technique. The problems of the invariant forms, associated with ordinary differential equations, is approached within the framework of the same formalism, thus displaying interesting relations with the Courant-Snyder invariant, introduced in the analysis of the motion of a charged particle along a transport channel, and with the Lewis-Riesenfeld invariant, introduced in the analysis of the evolution of a quantum harmonic oscillator with time-dependent frequency. Particular attention is devoted to the paraxial propagation of an electromagnetic wave through a non homogeneous medium and to the paraxial motion of a charged particle beam in a circular accelerator.

1. Introduction

Group theoretic methods have been successfully exploited in many fields of Physics. Just to quote a few examples, we recall the theory of Super Multiplet developed by Wigner [1] and Franzini-Radicati [2] accounting for the features of the nuclear spectra or the group-theoretic approach to the nuclear and particle spectroscopy, provided by the SU(3) theory of Elliott [3] and by the *eightfold way* of Gell'man and Ne'eman [4]. The formalism of the group theory has been applied also to classical and quantum optics, and, within a purely mathematical concern, to the theory of special functions and ordinary and partial differential equations.

Within the context of quantum optics, the use of group-theoretic methods for the analysis of the evolution of atoms and molecules, in which transitions are driven by electromagnetic fields, has received an increasing deal of attention in recent years. In their seminal papers, Elgin [5] and Hioe-Eberly [6] have stated the possibility of treating the evolution of a n-level quantum sustem interacting with a classical electromagnetic field using SU(n) as dynamical simmetry. The major results of their analysis are the discovery of previuosly unsuspected non-linear constants of motion, linked to the Casimir invariant of the group, and the possibility of casting the equations of motion, characterizing the evolution of the system, in terms of a generalized Bloch equation, thus allowing to visualize the evolution of the system as a precession of the *coherence vector* in a (n^2-1)-dimensional space. Such a visualization suggests interesting connections with methods, introduced in classical mechanics to treat rotations [7]. In fact, the functions introduced in the framework of the algebraic approach to ordering problems have been recognized as the Cayley-Klein parameters, which represent an alternative description to rotations in classical mechanics [8].

On the other hand, the use of operator techniques for the analysis of optical systems, originally suggested by Vander Lugt [9] has been later elaborated [10] within the context of a recognized structural similarity between classical optics and quantum mechanics, thus indicating the possibility of describing the propagation of a light wave along an optical system with the aid of a set of operators acting as canonical transformations in the framework of non relativistic quantum mechanics in two dimensions.

Finally, let us briefly comment on the contribution given by the theory of group representations within the attempt of understanding the theory of the special functions, which *at first sight appears as a chaotic collection of formulae*, from a single point of view [11]. It is worth stressing that the theory of special functions is strongly related to that of second-order ordinary differential equations, which have been studied with the aid of Lie-algebraic methods, and in fact Lie-groups were originally introduced to explore the deep reasons underlying the integrability by quadratures of ordinary differential equations [12].

On the other hand, the use of algebraic methods to deal with partial differential equations of evolution-type, which are the formal basis for many problems in Physics, is not new. Schrödinger applied the method of factorization to solve the time-dependent Schrodinger equation [13], and Miller showed that this method is equivalent to the representation theory of four Lie-algebras [14]. More recently, Lie-algebraic approach to the initial value and eigenvalue problems for certain special classes of partial differential operators, which occurr frequently in scientific literature, have been considered by Steinberg [15], thus providing detailed informations about these operators and explicit formulae for the solutions to the corresponding partial differential equations. Correspondently, in ref. [16], a general procedure for solving certain classes of partial differential equations, which combines Lie-algebraic methods with matrix techniques, has been proposed. In Sec. 2 the proposed method is presented with some details and the connections with other methods are pointed out. In particular, partial differential equations, which play an important role in quantum and classical optics are analysed in Sec. 3. A more detailed discussion is devoted to the parabolic wave equation, describing the paraxial propagation of an electromagnetic wave through a non-homogeneous medium exhibiting a parabolic dependence of the refractive index on the transverse coordinates [17] (Sec. 4). In Sec. 5, the matrix approach is applied to ordinary differential equation, thus providing a compact method to recover well-known results within the context of the theory of ordinary differentail equations [18] and yielding a new insight in the theory of the associated invariants. The analysis is limited to second order differential equations, since they play an important role in optics. The paraxial propagation of an e. m. wave through a non-homogeneous medium and the motion of a charged particle along a transport channel in the linear approximation are ruled by a second-order differential equation. For the discussion of higher order ordinary differential equations, the reader is addressed to ref. [18]. Furthermore, the algebraic approach suggests approximate methods, based for instance on a Magnus-type expansion [19], as complementary to the WKB method. In Sec. 6, the problem of the invariant forms associated with ordinary differential equations is analysed [20]. Finally, the analysis is applied to the motion of a charged particle in a circular accelerator (Sec. 7).

2. Matrix approach to partial differential equations

Let us consider the following Cauchy problem

$$\frac{\partial}{\partial\tau}\psi(\underline{x},\tau)=\hat{A}(\underline{x},\tau)\psi(\underline{x},\tau) \qquad \psi(\underline{x},0)=\phi(\underline{x}), \tag{2.1}$$

for the function ψ depending on τ and on the variables $\underline{x}=(x_1,\ldots\ldots,x_n)$. The function $\phi(\underline{x})$ specifying $\psi(\underline{x},\tau)$ at $\tau=0$ is assumed to be a sufficiently regular function. The differential operator $\hat{A}$ has, at present, an unspecified structure. Introducing the displacement operator $\hat{U}(\underline{x},\tau)$ defined as

$$\psi(\underline{x},\tau)=\hat{U}(\underline{x},\tau)\phi(\underline{x}), \tag{2.2}$$

Eq. (2.1) turns into the following

$$\frac{\partial}{\partial\tau}\hat{U}(\underline{x},\tau)=\hat{A}(\underline{x},\tau)\hat{U}(\underline{x},\tau) \qquad \hat{U}(\underline{x},0)=\hat{1}. \tag{2.3}$$

As is well known, many techniques have been developed in order to tackle with the above equation, as the Baker-Campbell-Hausdorff and Zassenhaus formulae, particularly useful for τ-independent $\hat{A}$ operators, or the Feynman-Dyson perturbative approach and the Magnus expansion, which deal with the ordering problem underlying Eq. (2.3) for τ-dependent $\hat{A}$ operators. A complete account of these method is provided by the review paper by Wilcox [21] and by ref. [22].

Within the present context of an algebraic view to the initial value problem, we will briefly discuss the technique proposed by Wei-Norman [23]. The method applies to operators $\hat{H}$, which can be expressed as a linear combinations of the generators $\hat{L}_i, i=1,\ldots,n$ of a n-dimensional Lie algebra G :

$$\hat{H}(\tau)=\sum_{i=1}^{n} a_i(\tau)\hat{L}_i, \tag{2.4}$$

with $a_i(\tau)$ τ-dependent complex functions. In that case, one can take advantages from a general theorem on Lie-group, which states the possibility of uniquely representing any element $\hat{U}$ in a small neighbourhood of the identity I in the group G associated with G as a product of elements

$$\hat{U}=\prod_{i=1}^{n} \exp(s_i\hat{L}_i), \tag{2.5}$$

with $\hat{L}_i, i=1,\ldots,n$ forming a basis for the algebra G and the complex numbers $s_i, i=1,\ldots,n$ lying in a small neighbourhood of zero. Consequently, since the evolution operator can be regarded as an element of the group G associated with G, it can be expressed as a product

$$\hat{U}(\tau)=\prod_{i=1}^{n} \exp[s_i(\tau)\hat{L}_i], \tag{2.6}$$

with s_i τ-dependent functions. Inserting (2.6) into the equation of motion (2.3) and exploiting the closure properties of the Lie-algebras, and the linear independence of the generators $\hat{L}_i$, we end up with the following system of equations:

$$a_i(\tau)=\sum_{j=1}^{n}\alpha_{ij}(s_1,\dots,s_n)\dot{s}_j(\tau) \qquad i=1,\dots,n \tag{2.7}$$

where α_{ij} are analytic functions of the s_j's, whose explicit functional form depends on the structure constants of the algebra involved in. The invertibility of the matrix $\underline{\underline{\alpha}}=(\alpha_{ij})$ is assured only in a neighbourhood of $\tau=0$, where the system (2.7) can be solved for the $\dot{s}_j$'s, thus getting

$$\dot{s}_i=\sum_{j=1}^{n}\eta_{ij}s_j(\tau). \tag{2.8}$$

As a consequence, the representation (2.6) cannot be assumed for all τ and $\underline{a}=(a_1,\dots,a_n)$. In ref. [18], it has been proved that the representation (2.6) is globally valid for all the solvable algebras and for the 3-dimensional Lie-algebra, whose generators $\hat{L}_1, \hat{L}_2$ and $\hat{L}_3$ obey the rules of commutation:

$$[\hat{L}_1,\hat{L}_2]=\lambda\hat{L}_2 \qquad [\hat{L}_1,\hat{L}_3]=-\lambda\hat{L}_3 \qquad [\hat{L}_2,\hat{L}_3]=2\delta\hat{L}_1, \tag{2.9}$$

where λ and δ are arbitrary numbers.
In that case, the system (2.8) specializes as

$$\begin{aligned}
\dot{s}_1(\tau)&=a_1-2\delta a_3 s_2 e^{\lambda s_1}\\
\dot{s}_2(\tau)&=a_2 e^{-\lambda s_1}+\lambda\delta a_3 s_2^2 e^{\lambda s_1}\\
\dot{s}_3(\tau)&=a_3 e^{\lambda s_1},
\end{aligned} \tag{2.10}$$

with initial conditions $s_i(0)=0, i=1,2,3$.

It can be easily proved that the system (2.10) is equivalent to the following Riccati equation for $\dot{s}_1(\tau)$:

$$\dot{u}-\lambda u^2+p(\tau)u+q(\tau)=0; \qquad u(0)=a_1(0) \quad u(\tau)=\dot{s}_1(\tau), \tag{2.11}$$

with

$$p(\tau)=-\frac{\dot{a}_3}{a_3}, \quad q(\tau)=-\dot{a}_1-\delta a_2 a_3+\lambda a_1^2+a_1\frac{\dot{a}_3}{a_3}. \tag{2.12}$$

It is worth noticing that the SU(2) and SU(1,1) algebras, which play a particularly important role in optics, are possible realizations of (2.9) with $\lambda=\delta=1$ and $\lambda=-\delta=1$ respectively. Just to quote a few examples, let us recall that the evolution of a two-level system interacting with an intense electromagnetic field is ruled by a Hamiltonian operator, which can be recognized to be an element of the SU(2) algebra. On the other hand, classical optics can be reformulated in terms of an algebraic language, the relevant group being SU(1,1) in the paraxial approximation.

More recently, Steinberg applied the Lie-algebraic methods to the initial value and eigenvalue problems for operators of the type

$$G=\left\{\hat{A}=a\frac{\partial^2}{\partial x^2}+bx\frac{\partial}{\partial x}+cx^2+\alpha\frac{\partial}{\partial x}+\beta x+\gamma; a,b,c,\alpha,\beta,\gamma \in \mathbf{C}\right\}, \tag{2.13}$$

with C denoting the set of complex numbers. It can be immediately realized that G is a Lie-algebra with respect to the Lie-product; namely

$$[\hat{A},\hat{B}]=\hat{A}\hat{B}-\hat{B}\hat{A}\ \in\ G, \tag{2.14}$$

for any $\hat{A}$ and $\hat{B}$ in G. In particular, choosing as basis for G the operators

$$\hat{L}_1=\frac{1}{2}\left(x\frac{\partial}{\partial x}+\frac{1}{2}\right)\quad \hat{L}_2=\frac{x^2}{2}\quad \hat{L}_3=\frac{1}{2}\frac{\partial^2}{\partial x^2}$$
$$\hat{L}_4=x\quad \hat{L}_5=\frac{\partial}{\partial x}\quad \hat{L}_6=\hat{I}, \tag{2.15}$$

it is easy to recognize G as the direct sum SU(1,1)⊕h(4), which is characterized by the rules of commutations

$$[\hat{L}_1,\hat{L}_2]=\hat{L}_2\quad [\hat{L}_1,\hat{L}_3]=-\hat{L}_3\quad [\hat{L}_2,\hat{L}_3]=-2\hat{L}_1\quad [\hat{L}_4,\hat{L}_5]=-\hat{L}_6$$
$$[\hat{L}_1,\hat{L}_4]=\frac{\hat{L}_4}{2}\quad [\hat{L}_1,\hat{L}_5]=-\frac{\hat{L}_5}{2}\quad [\hat{L}_2,\hat{L}_4]=0 \tag{2.16}$$
$$[\hat{L}_3,\hat{L}_4]=\hat{L}_5\quad [\hat{L}_2,\hat{L}_5]=-\hat{L}_4\quad [\hat{L}_3,\hat{L}_5]=0,$$

In that case, the system (2.7), relating the coefficients $a_1,\ldots,a_6$ of the linear combination (2.4) to the functions $s_1,\ldots,s_6$, which realize the representation (2.6) for the displacement operator $\hat{U}(\tau)$, can be solved for $\dot{s}_i(\tau)$'s, thus getting the following non linear system of differential equations:

$$\dot{s}_1 = a_1 + 2a_3 s_2 e^{s_1}$$
$$\dot{s}_2 = a_2 e^{-s_1} - a_3 s_2^2 e^{s_1}$$
$$\dot{s}_3 = a_3 e^{s_1} \tag{2.17}$$
$$\dot{s}_4 = a_4 e^{-\frac{s_1}{2}} + a_5 s_2 e^{\frac{s_1}{2}}$$
$$\dot{s}_5 = a_5(1 - s_2 s_3) e^{\frac{s_1}{2}} - a_4 s_3 e^{-\frac{s_1}{2}}$$
$$\dot{s}_6 = a_6 + s_4 \dot{s}_5 \qquad s_i(0) = 0 \quad i = 1,\dots,6.$$

The proof of the existence of the functions $s_i(\tau)$, which are "analytic (not necessarily single valued) functions for all τ and coefficients $a_1,\dots,a_6$ with the exception of certain isolated points" is one of the major results of the paper by Steinberg. Since Eq. (2.1) with $\hat{H}$-operators from the set (2.13) have many applications in physical problems, where it may occur in classical or non classical form, the detailed analysis carried out in ref. [15] allows to treat a wide class of problems within the context of a general and thorough formalism. Finally, let us notice that the system (2.17) can be separated into two sets of equations, one for the functions s_1, s_2 and s_3, which display the structure of (2.10) for $\lambda = 1$ and $\delta = -1$, and one for s_4, s_5 and s_6, which can be immediately integrated once the functional form of s_1, s_2 and s_3 is known. Such a separability of the system (2.17) is a consequence of that the algebra (2.13), as all the finite-dimensional algebras, can be decomposed into the direct sum of the radical and a semisimple subalgebra (Levi's theorem), which in the case we are considering correspond to h(4) and SU(1,1) respectively [22]. See also ref. [24], where the algebra SU(2) h(4) has been considered as well.

In ref. [16] it has been proposed an alternative procedure,

which in some sense combines the algebraic approach suggested by Wei-Norman with a matrix method, thus leading to a computationally more powerful technique. The proposed method reduces the problem of solving Eq. (2.1) or (2.3) into that of solving not a system of n non-linear ordinary differential equations, as it is within the context of the Wei-Norman type formalism, but a system of $m < n$ linear first order differential equations, with m being the dimensionality of the smallest representation of the algebra involved in.

In fact, hence any Lie-algebra G is isomorphic to a matrix algebra G', each generators $\hat{L}_i$ can be represented as an $m x m$ matrix. The dimensionality of the representation can be arbitrarily chosen; however, it is evident the convenience of choosing the smallest dimensionality representation. As a consequence, the linear combination (2.4) can be recast as an mxm matrix

$$\underline{\underline{H}}(\tau)=\begin{pmatrix} h_{11}(\tau) & \dots & h_{1m}(\tau) \\ \dots & \dots & \dots \\ h_{m1}(\tau) & \dots & h_{mm}(\tau) \end{pmatrix}, \tag{2.18}$$

with the matrix elements h_{ij} being in general complex functions of τ. Accordingly, within the context of the isomorphism $G \leftrightarrow G'$, Eq. (2.1) corresponds to the vector equation

$$\frac{\partial}{\partial\tau}\begin{pmatrix} \psi_1 \\ .. \\ \psi_m \end{pmatrix}(\tau)=\begin{pmatrix} h_{11} & \dots & h_{1m} \\ \dots & \dots & \dots \\ h_{m1} & \dots & h_{mm} \end{pmatrix}\begin{pmatrix} \psi_1 \\ .. \\ \psi_m \end{pmatrix}(\tau), \tag{2.19}$$

where the τ-dependent vector $\underline{\psi} \equiv (\psi_1(\tau),\dots,\psi_m(\tau))$ must be regarded as an element of a m-dimensional complex vector space, the meaning of $\underline{\psi}$ being specified by the problem under study.

As already noticed, the operator $\hat{U}$ belongs to the group G associated with G; therefore, it can be represented as an mxm matrix:

$$\underline{\underline{U}}(\tau)=\begin{pmatrix} u_{11}(\tau) & \dots & u_{1m}(\tau) \\ \dots & \dots & \dots \\ u_{m1}(\tau) & \dots & u_{mm}(\tau) \end{pmatrix}, \tag{2.20}$$

which acts on the vector $\underline{\psi}$ according to (2.2). The functions $u_{ij}(\tau)$ are defined by Eq. (2.3), which turns into m systems of first order ordinary differential equations for each column $(u_{j1},\dots,u_{jm})$ as

$$\frac{d}{d\tau}\begin{pmatrix} u_{j1} \\ \dots \\ u_{jm} \end{pmatrix}=\begin{pmatrix} h_{11} & \dots & h_{1m} \\ \dots & \dots & \dots \\ h_{m1} & \dots & h_{mm} \end{pmatrix}\begin{pmatrix} u_{j1} \\ \dots \\ u_{jm} \end{pmatrix}, \qquad j=1,\dots,m \tag{2.21}$$

with initial conditions $u_{jl}(0)=\delta_{jl}$. The systems (2.21) exhibit basically the same structure, the only difference being in the initial conditions, which assure $\hat{U}$ to be the unit matrix at $\tau=0$.

However, we need the explicit expression of $\hat{U}$ in the $\underline{x}$-representation in order to specify the functional form of ψ at any τ. To this end, let us assume according to the Wei-Norman uncoupling theorem the following form for $\hat{U}$:

$$\hat{U}(\tau)=\prod_{i=1}^{n}\exp[s_i(\tau)\hat{L}_i]. \tag{2.22}$$

Since each exponential operator $e^{s_i \hat{L}_i}$ is an element of the group G , it can be represented as an $m x m$ matrix, whose elements will be expressed in terms of the functions $s_i(\tau)$. Consequently, the product on the right hand of (2.22) can be given the form of an mxm matrix

$$\underline{\underline{U}}(\tau)=\begin{pmatrix} \xi_{11} & \dots & \xi_{1m} \\ \dots & \dots & \dots \\ \xi_{m1} & \dots & \xi_{mm} \end{pmatrix}, \tag{2.23}$$

whose elements ξ_{ij} are analytic functions of the s_i's. Equating the above expression to the matrix (2.20), we get a system of m algebraic equations, which can be solved for the s_i's, thus reducing the initial value problem, related to Eq. (2.1), to the solution of a set of m linear first-order differential equations.

As a conclusion, let us stress that it is crucial for applying the above described procedure in solving a partial differential equation of the type (2.1) to identify the algebra involved in. As a recipe, we can suggest to evaluate the commutators between the differential operators entering the equation under study, in order to find a close structure. So doing, it is possible to identify both the basis for the involved algebra and the relevant structure.

Finally, the lowest dimensionality matrix representation of the generators $\hat{L}_i$ can be obtained by noticing that

$$[\hat{L}_i, \hat{X}_k]=\sum_{j=1}^{m} L_{ij}\hat{X}_j \qquad k=1,\dots,m \quad i=1,\dots,n \tag{2.24}$$

with n and m denoting the dimensionality of the algebra and of the relevant matrix representation respectively. The operators $\hat{X}_1,\ldots,\hat{X}_m$ must be understood as the independent coordinates $x_1,\ldots,x_m$ for many-variables differential equation or as x and $\frac{\partial}{\partial x}$ for one-variable equation.

3. SU(2) and SU(1,1)- type equation.

The procedure described in the previous section will be now discussed within the framework of a specific example in which $\hat{H}$ is an element of the SU(1,1) or SU(2) algebra, i.e.

$$\hat{H}=\gamma(\tau)\hat{F}_3+\alpha(\tau)\hat{F}_++\beta(\tau)\hat{F}_-+\epsilon(\tau)\hat{1} \tag{3.1}$$

The operators $\hat{F}_3$ and $\hat{F}_\pm$ realize the above-quoted algebras according to the commutation rules

$$[\hat{F}_3,\hat{F}_\pm]=\pm\hat{F}_\pm \qquad [\hat{F}_+,\hat{F}_-]=2\delta\hat{F}_3 \tag{3.2}$$

and the parameter δ is allowed to take the values +1 and -1. Just to provide an example of the possible structures of $\hat{H}$ in terms of differential operators, we notice that

$$\hat{F}_3=\frac{1}{2}\left(y\frac{\partial}{\partial y}+x\frac{\partial}{\partial x}\right) \qquad \hat{F}_+=y\frac{\partial}{\partial x} \qquad \hat{F}_-=x\frac{\partial}{\partial y} \tag{3.3}$$

realize a SU(2) algebra. Correspondently, the operators (see also eq. (2.15))

$$\hat{F}_3=\frac{1}{2}\left(x\frac{\partial}{\partial x}+\frac{1}{2}\right) \qquad \hat{F}_+=\frac{x^2}{2} \qquad \hat{F}_-=\frac{1}{2}\frac{\partial^2}{\partial x^2} \tag{3.4}$$

or

$$\hat{F}_3=\frac{1}{2}\left(x\frac{\partial}{\partial x}+y\frac{\partial}{\partial y}+1\right) \qquad \hat{F}_+=xy \qquad \hat{F}_-=\frac{\partial^2}{\partial x\partial y} \tag{3.5}$$

represent two possible realizations of the SU(1,1) algebra.

It is worth stressing that on exploiting the Bargmann representation of the creation-annihilation operators i. e.

$$\hat{a}^+=x \qquad \hat{a}=\frac{\partial}{\partial x} \tag{3.6}$$

the realizations (3.3-5) can be obtained according to the Schwinger prescription. The relevance of the operators (3.1) in quantum mechanics, for both hermitian and non-hermitian dynamics, is immediately recognized.

As is well known, the operators (3.2) can be represented by the $2x2$ matrices

$$\hat{F}_3=\frac{1}{2}\begin{pmatrix}-1 & 0\\ 0 & 1\end{pmatrix} \qquad \hat{F}_+=\begin{pmatrix}0 & 0\\ \delta & 0\end{pmatrix} \qquad \hat{F}_-=\begin{pmatrix}0 & 1\\ 0 & 0\end{pmatrix} \tag{3.7}$$

which yields for the operator $\hat{A}$ the following matrix form

$$\underline{\underline{H}} = \begin{pmatrix} -\frac{\gamma}{2}+\epsilon & \beta \\ \delta\alpha & \frac{\gamma}{2}+\epsilon \end{pmatrix} \tag{3.8}$$

where unit matrix has been introduced as required by (3.1). The evolution of the displacement matrix (*)

$$\underline{\underline{U}} = \begin{pmatrix} H & F \\ G & L \end{pmatrix} \tag{3.9}$$

is ruled by the set of first order differential equations:

$$\begin{aligned} &\dot{H} = \left(-\frac{\gamma}{2}+\epsilon\right)H + \beta G \\ &\dot{G} = \delta\alpha H + \left(\frac{\gamma}{2}+\epsilon\right)G, \qquad H(0)=1 \quad G(0)=0 \\ &\dot{F} = \left(-\frac{\gamma}{2}+\epsilon\right)F + \beta L \\ &\dot{L} = \delta\alpha F + \left(\frac{\gamma}{2}+\epsilon\right)L, \qquad F(0)=0 \quad L(0)=1. \end{aligned} \tag{3.10}$$

(*) The notation in term of the functions H, F, G and L has been introduced just to share the language of refs. [25,26] where the present procedure has been applied to analyse the SU(2) and SU(1,1)-type evolution within the context of a hermitian and non-hermitian dynamics.

The functions H and F satisfy the same second order differential equation

$$\ddot{y}(\tau)+p(\tau)\dot{y}(\tau)+q(\tau)y(\tau)=0 \tag{3.11}$$

with

$$p(\tau)=-\frac{\dot{\beta}}{\beta}-2\epsilon$$
$$q(\tau)=-\left(\frac{\gamma}{2}-\epsilon\right)\frac{\dot{\beta}}{\beta}+\frac{\dot{\gamma}}{2}-\dot{\epsilon}-\delta\alpha\beta-\frac{\gamma^2}{4}+\epsilon^2 \tag{3.12}$$

and initial conditions

$$H(0)=1 \quad \dot{H}(0)=-\frac{\gamma(0)}{2}+\epsilon(0) \;, \quad F(0)=0 \quad \dot{F}(0)=\beta(0) \tag{3.13}$$

Correspondently, the functions G and L obey the same equation (3.11) with

$$p(\tau)=-\frac{\dot{\alpha}}{\alpha}-2\epsilon \qquad q(\tau)=\frac{\dot{\alpha}}{\alpha}\left(\frac{\gamma}{2}+\epsilon\right)-\delta\alpha\beta-\frac{\dot{\gamma}}{2}-\dot{\epsilon}-\frac{\gamma^2}{4}+\epsilon^2 \tag{3.14}$$

4. Paraxial propagation of an e.m. wave through a non homogeneous medium.

The propagation of a linearly polarized e.m. wave through a non-homogeneous medium is ruled by the scalar Helmholtz equation

$$\nabla^2 u(x,y,z) + k^2 n^2(x,y,z) u(x,y,z) = 0, \tag{4.1}$$

with k denoting the wave number in the vacuum and $n(x,y,z)$ the refractive index of the medium.
In the paraxial approximation the refractive index can be expressed as

$$n(x,y,z) \simeq n_0(z) - \frac{1}{2} n_{2x}(z) x^2 - \frac{1}{2} n_{2y}(z) y^2, \tag{4.2}$$

where z is the coordinate along the direction of propagation and x and y the coordinates in the transverse plane.
Assuming for $u(x,y,z)$ the form of a quasi-plane wave:

$$u(x,y,z) = \phi(x,y,z) \exp\left\{ -ik \int_{z_0}^{z} n_0(z') dz' \right\}, \tag{4.3}$$

in the hypothesis of slowly varying amplitude we get for ϕ the following equation

$$\nabla_\perp^2 \phi - 2ikn_0(z) \frac{\partial}{\partial z} \phi + \left[k^2 (n^2(x,y,z) - n_0^2(z)) - ik \frac{d}{dz} n_0(z) \right] \phi = 0, \tag{4.4}$$

where $\nabla_\perp^2$ is the transverse laplacian. A further simplification is possible, by introducing the function $\chi(x,y,z)$ as

$$\phi(x,y,z)=\frac{1}{\sqrt{n_0(z)}}\chi(x,y,z), \tag{4.5}$$

which yields the equation

$$\frac{i}{k}\frac{\partial}{\partial z}\chi=\frac{1}{2}\left[\frac{1}{k^2 n_0(z)}\nabla_\perp^2+\frac{1}{n_0(z)}(n^2(x,y,z)-n_0^2(z))\right]\chi. \tag{4.6}$$

It is immediate to recognize the structural similarity of the above equation with the Schrödinger equation, governing the motion of a particle with variable mass $k^2 n_0(z)$ acted by the potential $\{n^2(x,y,z)-n_0^2(z)\}/n_0(z)$. Within this concern, the longitudinal coordinate z and the reduced wavelength $\lambda\!\!\!^{-}=\frac{1}{k}$ replace the time and the reduced Planck constant respectively [27]. Inserting the expression (4.2) for the refractive index into (4.6) specializes the potential of eq. (4.6) as that of a harmonic oscillator, i. e.

$$\frac{i}{k}\frac{\partial}{\partial z}\chi(x,y,z)=\frac{1}{2}\{\frac{1}{k^2 n_0(z)}\nabla_\perp^2-[n_{2x}(z)x^2+n_{2y}(z)y^2]\}\chi(x,y,z). \tag{4.7}$$

The structure of the above equation suggests to separate the dependence of χ on the transverse coordinates as

$$\chi(x,y,z)=\psi_x(x,z)\psi_y(y,z), \tag{4.8}$$

where ψ_x and ψ_y are easily checked to obey the equation

$$\frac{i}{k}\frac{\partial}{\partial z}\psi(\eta,z)=\frac{1}{2}\left\{\frac{1}{k^2 n_0(z)}\frac{\partial^2}{\partial\eta^2}-n_2(z)\eta^2\right\}\psi(\eta,z), \tag{4.9}$$

with n_2 and η denoting n_{2x} or n_{2y} and x or y respectively, according to whether ψ stands for ψ_x or ψ_y .

Equation (4.9) exhibits the same structure of Eq. (2.1) with the operator $\hat{H}$

$$\hat{H}(z)=-\frac{i}{2}\left\{\frac{1}{kn_0(z)}\frac{\partial^2}{\partial\eta^2}-kn_2(z)\eta^2\right\}, \tag{4.10}$$

being immediately recognized as a linear combination of the operators

$$\hat{K}_3=\frac{1}{2}\left(\eta\frac{\partial}{\partial\eta}+\frac{1}{2}\right)\qquad \hat{K}_+=\frac{i}{2}\eta^2\qquad \hat{K}_-=-\frac{i}{2}\frac{\partial^2}{\partial\eta^2}, \tag{4.11}$$

which according to the relations of commutation

$$[\hat{K}_3,\hat{K}_\pm]=\pm\hat{K}_\pm\qquad [\hat{K}_+,\hat{K}_-]=-2\hat{K}_3 \tag{4.12}$$

realize a SU(1,1) algebra.

As noticed in the previous section, the operators (4.11) can be represented by the $2x2$ matrices

$$\hat{K}_3=\frac{1}{2}\begin{pmatrix}-1 & 0\\ 0 & 1\end{pmatrix}\quad \hat{K}_+=\begin{pmatrix}0 & 0\\ -1 & 0\end{pmatrix}\quad \hat{K}_-=\begin{pmatrix}0 & 1\\ 0 & 0\end{pmatrix}, \tag{4.13}$$

thus allowing to rewrite Eq. (4.9) in the matrix form:

$$\frac{\partial}{\partial z}\begin{pmatrix}\psi_1\\ \psi_2\end{pmatrix}(z)=\begin{pmatrix}0 & \dfrac{1}{kn_0(z)}\\ -kn_2(z) & 0\end{pmatrix}\begin{pmatrix}\psi_1\\ \psi_2\end{pmatrix}(z). \tag{4.14}$$

The meaning of the components of the vector $\underline{\psi}=(\psi_1,\psi_2)$ can be directly inferred from eq. (4.14), which suggests to regard (ψ_1,ψ_2) as the ray-coordinates, usually adopted in paraxial optics to describe the evolution of a meridional ray along an optical system [28]. In particular, $\psi_1(z)$ can be understood as the height r of the ray at z and $\psi_2(z)$ as the reduced slope $r'(z)$; explicitly

$$\psi_1(z)=r(z) \qquad \psi_2(z)=kn_0\frac{d}{dz}r=r'(z). \tag{4.15}$$

Consequently, Eqs. (4.14) are nothing but the ray equation [28] in the paraxial approximation, rewritten in form of a system of two first-order differential equations.

According to the formalism of ray-optics, let us write the solution to eq. (4.14) in the form

$$\underline{\psi}(z)=\begin{pmatrix}A(z,z_0) & B(z,z_0)\\ C(z,z_0) & D(z,z_0)\end{pmatrix}\underline{\psi}(z_0), \tag{4.16}$$

where the $ABCD$ matrix plays the role of the displacement operator introduced within the context of the general treatment of Sec. 2. The matrix elements obey the equations

$$\frac{d}{dz}A = \frac{1}{kn_0(z)}C$$

(4.17a)

$$\frac{d}{dz}C = -kn_2(z)A \qquad A(z_0) = 1, C(z_0) = 0$$

$$\frac{d}{dz}B = \frac{1}{kn_0(z)}D$$

(4.17b)

$$\frac{d}{dz}D = -kn_2(z)B \qquad D(z_0) = 1, B(z_0) = 0$$

which suggest to recast the matrix (4.16) in the more compact form

$$\underline{\underline{U}}(z) = \begin{pmatrix} A & B \\ A' & B' \end{pmatrix}, \tag{4.18}$$

where A' and B' are defined as

$$A' = kn_0(z)\frac{d}{dz}A \qquad B' = kn_0(z)\frac{d}{dz}B. \tag{4.19}$$

The functions A and B satisfy the ray-equation

$$\frac{d}{dz}\left(kn_0(z)\frac{d}{dz}y\right) + kn_2(z)y = 0, \tag{4.20}$$

with initial conditions

$$A(z_0, z_0) = 1 \qquad A'(z_0, z_0) = 0 \ , \ B(z_0, z_0) = 0 \qquad B'(z_0, z_0) = 1 . \quad (4.21)$$

As a consequence, according to (4.19), the propagation matrix can be understood as completely specified by two linearly independent solutions to the ray-equation (4.20) corresponding to complementary initial conditions as given by (4.21). In fact, A represents a ray, which enters the optical system parallel to the optical axis and with a unitary height. Conversely B describes the evolution of a ray which initially has a vanishing height and a unitary reduced slope.

In the case where n_0 and n_2 do not change with z, we get the well-known result:

$$\begin{pmatrix} r \\ r' \end{pmatrix}(z) = \begin{pmatrix} \cos\sqrt{\dfrac{n_2}{n_0}}z & \dfrac{1}{\sqrt{n_0 n_2}}\sin\sqrt{\dfrac{n_2}{n_0}}z \\ -\sqrt{n_0 n_2}\sin\sqrt{\dfrac{n_2}{n_0}}z & \cos\sqrt{\dfrac{n_2}{n_0}}z \end{pmatrix} \begin{pmatrix} r_0 \\ r'_0 \end{pmatrix} \qquad (4.23)$$

for $n_2 > 0$.

In conclusion, the matrix approach to the Helmholtz equation leads to a description of the propagation in terms of geometrical optics.

Let us turn now to the evolution of a field distribution. According to the procedure of Sec. 2, let us represent the propagation operator $\hat{U}(\eta, z)$, which in the coordinate representation corresponds to the previously introduced matrix $ABCD$, as the product

$$\hat{U}(\eta, z) = \exp(2h\hat{K}_3)\exp(g\hat{K}_+)\exp(f\hat{K}_-) \qquad (4.24)$$

As already noticed, according to (3.7), the exponential operators entering the above expression can be represented by $2x2$ matrices, which within the present context have a specific meaning. In fact, $e^{f\hat{K}_-}$ writes in the following matrix form

$$e^{f\hat{K}_-} = \begin{pmatrix} 1 & f \\ 0 & 1 \end{pmatrix}, \tag{4.25}$$

which can be immediately recognized as the ray-matrix relevant to the propagation through a straight section of length f. Correspondently, the operator

$$e^{g\hat{K}_+} = \begin{pmatrix} 1 & 0 \\ -g & 1 \end{pmatrix} \tag{4.26}$$

gives the ray-matrix of a thin lens with focus $\frac{1}{g}$. Finally

$$e^{2h\hat{K}_3} = \begin{pmatrix} e^{-h} & 0 \\ 0 & e^{h} \end{pmatrix} \tag{4.27}$$

can be understood as the ray-matrix corresponding to a beam-expander with magnification $M = e^{-h}$. It is interesting to notice that the propagation through an optical system described by the ray-matrix $e^{2h\hat{K}_3}$ can be realized by appropriately combining the ray-matrices relevant to the free-propagation and to the transmission through a thin lens. In fact, the optical system schetcked in Fig. 1 is described by the ray-matrix $\begin{pmatrix} M & 0 \\ 0 & \frac{1}{M} \end{pmatrix}$ with the magnification M being given by $M = -\frac{f_2}{f_1}$. Consequently, one can express the group element $e^{2h\hat{K}_3}$ as a product of elements $e^{\alpha\hat{K}_+}$ and $e^{\beta\hat{K}_-}$, according to

$$e^{2h\hat{K}_3} = e^{f_1\hat{K}_-} e^{f_1\hat{K}_+} e^{(f_1+f_2)\hat{K}_-} e^{f_2\hat{K}_+} e^{f_2\hat{K}_-}, \tag{4.28}$$

with $f_1(\tau)$ and $f_2(\tau)$ being arbitrarily chosen to satisfy the condition

$$e^{-h} = -\frac{f_2}{f_1} = A. \tag{4.29}$$

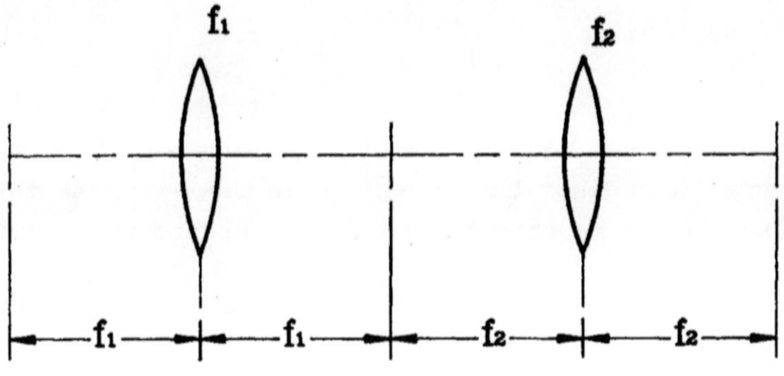

Fig. 1. Optical system with magnification M

Roughly speaking, one can say that the operator $\hat{U}$ takes into account the effect of the propagation through an optical system by successive adjustment of the spot-size and the wave-front curvature of the optical beam realized by an appropriate set of lenses and free-propagations, which are the basic optical elements in the paraxial approximation.

Finally, the operator $\hat{U}$ can be given the following expression

$$\underline{\underline{U}}(z,z_0)=\begin{pmatrix} e^{-h} & fe^{-h} \\ -ge^{h} & (1-fg)e^{h} \end{pmatrix}, \qquad (4.30)$$

which immediately provides the set of algebraic equations defining the functions h, g and f:

$$e^{-h}=A \qquad fe^{-h}=B \qquad ge^{h}=-A' \qquad (1-fg)e^{h}=B'. \qquad (4.31)$$

Solving for h, g and f, we end up with

$$h=-\ln A, \quad g=-\frac{k}{2}n_0(z)\frac{d}{dz}A^2, \quad f=\frac{B}{A}. \qquad (4.32)$$

Let us specify now the explicit effect of the operator $\hat{U}(\eta,z)$ on a field distribution $\psi(\eta,z_0)$ in the η-representation. According to the representation (4.24) and to the relations (4.31) we get

$$\psi(\eta,z)=\sqrt{\frac{i}{2\pi B}}\int \exp\{-\frac{i}{2B}(\eta^2 D+\eta'^2 A-2\eta\eta'\}\psi(\eta',z_0)d\eta' \qquad (4.33)$$

which can be immediately recognized as the Huygens-Kirchoff integral.

As a conclusion to this section, it is worth stressing that the matrix approach to the parabolic-wave equation (4.9), which can be regarded as an evolution equation within the context of wave-optics, allows to turn the problem into that of solving the paraxial ray-equation, which is substantially the counterpart of the parabolic wave-equation within the context of the Hamiltonian approach to geometrical optics.

5. Ordinary differential equations: matrix approach.

The formalism so far developed consists essentially in reducing the problem of solving an evolution-type partial differential equation, involving a 'Hamiltonian' operator belonging to a finite-dimensional Lie-algebra into that of solving a system af first-order ordinary differential equations, which is determined by the algebraic structure involved in the problem under study.
In this section, we show that the same procedure might provide a useful and compact tool of analysis for ordinary differential equations of arbitrary order [18,20].

For illustrative purposes, we consider the homogeneous second-order differential equation

$$\ddot{y} + p(\tau)\dot{y} + q(\tau)y = 0, \tag{5.1}$$

with initial conditions

$$y(0) = y_0 \qquad \dot{y}(0) = \dot{y}_0. \tag{5.2}$$

The dot denotes derivative with respect to τ, and the functions p and q are sufficiently regular functions of τ.

As is well known, Eq. (5.1) can be cast in the form of a system of two first-order differential equations according to

$$\frac{d}{d\tau}\begin{pmatrix} y_1 \\ y_2 \end{pmatrix}(\tau)=\begin{pmatrix} 0 & 1 \\ -q(\tau) & -p(\tau) \end{pmatrix}\begin{pmatrix} y_1 \\ y_2 \end{pmatrix}(\tau), \tag{5.3}$$

after identifying y_1 and y_2 with y and $\dot{y}$ respectively.

Equation (5.3) can be regarded as a vector equation in a two-dimensional vector space, whose solution can be formally written as

$$\underline{y}(\tau)=\underline{\underline{U}}(\tau)\underline{y}(0), \tag{5.4}$$

with $\underline{y}(0)$ denoting the vector containing the initial conditions.

$$\underline{y}(0)=\begin{pmatrix} y_0 \\ \dot{y}_0 \end{pmatrix}, \tag{5.5}$$

and the displacement matrix $\underline{\underline{U}}(\tau)$ summarizing all the features of the evolution. Just to share a common language with ray-optics, let us express $\underline{\underline{U}}(\tau)$ as

$$\underline{\underline{U}}(\tau)=\begin{pmatrix} A(\tau) & B(\tau) \\ C(\tau) & D(\tau) \end{pmatrix}, \tag{5.6}$$

with the functions A, B, C and D obeying the set of equation

$$\dot{A}=C$$
$$\dot{C}=-qA-pC \quad , \quad A(0)=1 \quad C(0)=0$$
$$\tag{5.7}$$
$$\dot{B}=D$$
$$\dot{D}=-qB-pD \quad , \quad B(0)=0 \quad D(0)=1.$$

It is immediate to recognize the possibility of recasting $\underline{\underline{U}}$ in the form

$$\underline{\underline{U}}(\tau)=\begin{pmatrix} A(\tau) & B(\tau) \\ \dot{A}(\tau) & \dot{B}(\tau) \end{pmatrix}, \tag{5.8}$$

with A and B being easily realized as two linearly independent solutions to the original equation (5.1) corresponding to the complementary initial conditions

$$A(0)=1, \dot{A}(0)=0 \qquad B(0)=0, \dot{B}(0)=1. \tag{5.9}$$

As a consequence, the matrix $\underline{\underline{U}}$ stands out as the wronskian of the cos-type (function A) and sin-type (function B) solutions to Eq. (5.1).
It can be easily proved that the determinant of $\underline{\underline{U}}$, $W(\tau)\equiv det\underline{\underline{U}}(\tau)$, obeys the equation

$$\frac{d}{d\tau}W(\tau)=-pW(\tau), \tag{5.10}$$

which immediately provides for $W(\tau)$ the expression

$$W(\tau) = \exp\left\{ -\int_0^\tau p(\tau')d\tau' \right\}, \tag{5.11}$$

according to the value of W at $\tau = 0$, i. e. $W(0) = 1$, thus recovering the Abel identity [29].

A similar approach to a non-homogeneous second-order differential equation

$$\ddot{y} + p(\tau)\dot{y} + q(\tau)y = \phi(\tau) \tag{5.12}$$

leads to the vector equation

$$\frac{d}{d\tau}\underline{y}(\tau) = \underline{\underline{H}}(\tau)\underline{y}(\tau) + \underline{\phi}(\tau), \tag{5.13}$$

where $\underline{\underline{H}}$ is the $2x2$ matrix introduced in (5.3) and $\underline{\phi}$ denotes the vector

$$\underline{\phi}(\tau) = \begin{pmatrix} 0 \\ \phi & (\tau) \end{pmatrix}. \tag{5.14}$$

According to the usual procedure, let us write the solution to (5.14) as

$$\underline{y}(\tau) = \underline{y}_0(\tau) + \underline{Y}(\tau), \tag{5.15}$$

with $\underline{y}_0(\tau)$ denoting a general solution to the homogeneous equation corresponding to (5.13), i. e.

$$\frac{d}{d\tau}\underline{y}_0(\tau)=\underline{\underline{H}}(\tau)\underline{y}_0(\tau) \tag{5.16}$$

and $\underline{Y}(\tau)$ being a particular solution to (5.13).

Accordingly, let us assume for $\underline{Y}(\tau)$ the following equation

$$\underline{Y}(\tau)=\underline{\underline{U}}(\tau)\underline{\xi}(\tau), \tag{5.18}$$

where $\underline{\underline{U}}_0(\tau)$ is the matrix evolution relevant to (5.16) and the vector $\underline{\xi}(\tau)$ is defined by the equation

$$\frac{d}{d\tau}\underline{\xi}(\tau)=\underline{\underline{U}}^{-1}\underline{\phi}(\tau). \tag{5.19}$$

The solution to the above equation is immediately written as

$$\underline{\xi}(\tau)=\int_0^\tau \underline{\underline{U}}^{-1}(\tau)\underline{\phi}(\tau), \tag{5.20}$$

which, by taking the explicit expression of the matrix $\underline{\underline{U}}^{-1}$ and the vector $\underline{\phi}$, specializes into

$$\underline{\xi}(\tau)=\int_0^\tau \frac{1}{W(\tau)}\begin{pmatrix}- & B & (\tau') \\ & A & (\tau')\end{pmatrix}\phi(\tau')d\tau', \tag{5.21}$$

with A and B being the elements of the evolution matrix relevant to the homogeneous equation.
Consequently, according to (5.15) one can express the solution to (5.12) as

$$y(\tau)=A(\tau)y_0+B(\tau)\dot{y}_0+B(\tau)\int_0^\tau \frac{1}{W(\tau')}A(\tau')\phi(\tau')d\tau'-A(\tau)\int_0^\tau \frac{1}{W(\tau')}B(\tau')\phi(\tau')d\tau', \tag{5.22}$$

that is in terms of the non-homogeneous term $\phi(\tau)$ and the cos-type and sin-type solutions to the corresponding homogeneous equation.

Let us turn now to Eq. (5.1) in order to identify the underlying algebaic structure. It is well known indeed that a $2x2$ matrix can be regarded as an element of the group U(2) or the symplectic group Sp(2,R)(*). Accordingly, the 2x2 matrix representation of the generators of the symplectic algebra sp(2,R), as given in (4.13), allows to express the matrix $\underline{\underline{H}}$ as the following linear cmbination

$$\underline{\underline{H}}(\tau)=-p(\tau)\underline{\underline{K}}_3+q(\tau)\underline{\underline{K}}_+ +\underline{\underline{K}}_- -\frac{1}{2}p(\tau)\underline{\underline{I}}, \tag{5.23}$$

(*) It is worth noticing that the algebras SU(2) and SU(1,1), characterized by the commutation rules (3.2) with $\delta=1$ and $\delta=-1$ respectively, are quite equivalent. Redefining indeed the generators of the two algebras to a set of $\{\hat{L}_\alpha\}_{\alpha=\pm,3}$, where the sign of $\hat{L}_+$ or $\hat{L}_-$ is changed with respect to $\hat{F}_+$ or $\hat{F}_-$, we can recover the other algebra. In cartesian components, this is a complex linear combination, which essentially performs the Weyl trick on SO(2,1)↔ SO(3).

where the matrices $\underline{\underline{K}}_3, \underline{\underline{K}}_\pm$ are characterized by the commutaion relations (4.12).

Consequently, according to the Wei-Norman theorem, the evolution matrix associated with Eq. (5.1) can be expressed as the following product

$$\underline{\underline{U}}(\tau) = \exp\{2h(\tau)\underline{\underline{K}}_3\}\exp\{g(\tau)\underline{\underline{K}}_+\}\exp\{f(\tau)\underline{\underline{K}}_-\}\exp\{d(\tau)\underline{\underline{I}}\}, \qquad (5.24)$$

which provides for $\underline{\underline{U}}$ the explicit representation

$$\underline{\underline{U}}(\tau) = \begin{pmatrix} e^{-h+d} & fe^{-h+d} \\ -ge^{h+d} & (1-fg)e^{h+d} \end{pmatrix}. \qquad (5.25)$$

The ordering functions h, f, g and d are defined by the set of differential equations

$$\begin{aligned} \dot{h} &= -\frac{p}{2} + ge^{2h} \\ \dot{g} &= qe^{-2h} - g^2e^{2h} \\ \dot{f} &= e^{2h} \\ \dot{d} &= -\frac{p}{2}, \qquad h(0) = g(0) = f(0) = d(0) = 0 \end{aligned} \qquad (5.26)$$

which can be easily combined to get the following Riccati equation for $\dot{h}$:

$$\ddot{h} - \dot{h}^2 - \omega(\tau) = 0, \qquad (5.27)$$

with initial conditions $\dot{h}(0) = -\frac{p(0)}{2}$. The function $\omega(\tau)$, explicitly given by

$$\omega(\tau) = q - \frac{1}{2}\dot{p} - \frac{1}{4}p^2 \tag{5.28}$$

can be immediately recognized as the *invariant* [29] associated with any linear second-order differential equation of the type (5.1), which by means of the well known substitution

$$y(\tau) = v(\tau)\exp\left\{-\frac{1}{2}\int_0^\tau p(\tau')d\tau'\right\}, \tag{5.29}$$

is transformed into the *normal form* (*)

$$v(\tau) + \omega(\tau)v(\tau) = 0. \tag{5.30}$$

Let us consider with some details the above equation, for it desplays interesting features. The evolution matrix $\underline{\underline{U}}(\tau)$ associated with Eq. (5.30), which can be rewritten in vector form as

$$\frac{d}{d\tau}\underline{v}(\tau) = \underline{\underline{H}}(\tau)\underline{v}(\tau), \tag{5.31}$$

(*) Let us notice that the Riccati equation (5.27) for $\dot{h}$ can be rewritten in the form (5.30), by setting $\dot{h} = -\frac{\dot{v}}{v}$.

with $\underline{\underline{H}}(\tau)$ being the matrix

$$\underline{\underline{H}}(\tau)=\begin{pmatrix} 0 & 1 \\ -\omega(\tau) & 0 \end{pmatrix} \tag{5.32}$$

has a unit determinant, as can be easily inferred from the general relation (5.11). In the two-dimensional case,we are dealing with, the unit value of the determinant assures $\underline{\underline{U}}$ to satisfy the matrix equation

$$\underline{\underline{U}}^T \underline{\underline{J}}\,\underline{\underline{U}}=\underline{\underline{J}}, \tag{5.33}$$

with $\underline{\underline{J}}$ denoting the matrix

$$\underline{\underline{J}}=\begin{pmatrix} 0 & -1 \\ 1 & 0 \end{pmatrix}. \tag{5.34}$$

Equation (5.33) is the condition that $\underline{\underline{U}}$ be a symplectic matrix. Accordingly the matrix $\underline{\underline{H}}$ satisfies the condition (*)

(*) It is worth stressing that the relation (5.33) can be regarded as the 'classical' counterpart of the unitary condition, satisfied by the quantum evolution operator. In fact, in terms of the corresponding symplectic conjugate $\underline{\underline{\tilde{U}}}$, defined according to (5.37), the relation (5.33) can be rewritten as

$$\underline{\underline{U}}\,\underline{\underline{\tilde{U}}}=\underline{\underline{\tilde{U}}}\,\underline{\underline{U}}=\underline{\underline{I}}$$

Similarly, the relation (5.36) plays, within the context of classical mechanics, the same role played by the hermiticity condition, which the quantum Hamiltonian operator is required to satisfy.

$$\underline{\underline{H}}^T \underline{\underline{J}} + \underline{\underline{J}}\,\underline{\underline{H}} = 0, \tag{5.35}$$

which implies the following relation between $\underline{\underline{H}}$ and the corresponding symplectic conjugate $\underline{\underline{\tilde{H}}}$:

$$\underline{\underline{H}} = -\underline{\underline{\tilde{H}}}, \tag{5.36}$$

with $\underline{\underline{\tilde{H}}}$ being defined as

$$\underline{\underline{\tilde{H}}} = -\underline{\underline{J}}\,\underline{\underline{H}}^T \underline{\underline{J}}. \tag{5.37}$$

The relation (5.33) or (5.35) suggests the possibility to associate Eq. (5.30) or equivalently (5.31) with a 'dynamical' problem ruled by the Hamiltonian

$$H(\tau) = \frac{1}{2}\underline{\upsilon}^T(\tau)\underline{\underline{J}}\,\underline{\underline{H}}\,\underline{\upsilon}(\tau), \tag{5.38}$$

where $\underline{\upsilon}^T$ denotes the transpose of $\underline{\upsilon}$ and $\underline{\underline{H}}$ is given in (5.32). Consequently, $\underline{\underline{H}}$ specializes as

$$H(\tau) = \frac{1}{2}\upsilon_2^2(\tau) + \frac{1}{2}I(\tau)\upsilon_1^2(\tau), \tag{5.39}$$

with the components of the vector $\underline{\upsilon}$, υ_1 and υ_2, playing, within the present context, the role of conjugate variables. Their evolution is indeed determined by the Hamilton's equations of motion

$$\dot{v}_1 = \frac{\partial}{\partial v_2} H$$

$$\dot{v}_2 = -\frac{\partial}{\partial v_1} H, \tag{5.40}$$

which immediately leads to (5.31).
Before concluding this section, let us notice that the general form of a second order differential equation, which can be associated with a Hamiltonian dynamics, is provided by the ray-equation (4.20). It is well known indeed that the paraxial ray-equation is derived from the quadratic Hamiltonian

$$H(z) = \frac{1}{2n_0(z)} r'^2 + \frac{1}{2} n_2(z) r^2, \tag{5.41}$$

with r and r' being the ray-coordinates (see Sec.4). Consequently, the corresponding ray-matrix satisfies the relation (5.33), which in optics language is usually reported as 'lens equation'.

Finally, it is worth stressing that the transformation

$$y(\tau) = v(\tau) a(\tau) \exp\left\{ -\int_0^\tau \frac{b(\tau')}{a(\tau')} d\tau' \right\} \tag{5.42}$$

into the 'normal' form

$$\frac{d}{d\tau}[n_0(\tau)\dot{v}] + n_2(\tau) v = 0, \tag{5.43}$$

where

$$n_0(\tau) = a^2(\tau)\exp\left\{-\int_0^\tau \frac{b(\tau')}{a(\tau')}d\tau'\right\},$$

$$n_2(\tau) = (\ddot{a} - \dot{b} + c)\exp\left\{-\int_0^\tau \frac{b(\tau')}{a(\tau')}d\tau'\right\}. \tag{5.44}$$

In the forthcoming section, Eq. (5.43) will be considered within the context of the invariant forms, which can be associated with ordinary differential equations. Equations of the type (5.43) are particularly interesting: the paraxial propagation of an optical beam through a non-homogeneous medium (Sec. 4) and the paraxial motion of a charged particle along a transport channel (Sec. 7) are ruled by equations formally similar to (5.43).

Finally, let us stress the possibility of associating any ordinary differential equation of order $2n$ with a dynamical problem with n degree of freedom. A $2n$-order differential equation, which in vector form writes

$$\frac{d}{d\tau}\underline{y}(\tau) = \underline{\underline{H}}(\tau)\underline{y}(\tau), \tag{5.45}$$

with $\underline{y}$ being a $2n$-component vector and $\underline{\underline{H}}$ a $2nx2n$, can be cast into a 'normal' form by means of an appropriate transformation. In fact, the vector $\underline{v}$ defined as

$$\underline{v}(\tau) = \underline{\underline{T}}(\tau)\underline{y}(\tau), \tag{5.46}$$

with $\underline{\underline{T}}$ arbitrary $2n\times 2n$ matrix, satisfies the equation

$$\frac{d}{d\tau}\underline{v}(\tau)=\underline{\underline{H}}'(\tau)\underline{v}(\tau), \tag{5.47}$$

with $\underline{\underline{H}}'$ being given by

$$\underline{\underline{H}}'=\dot{\underline{\underline{T}}}\,\underline{\underline{T}}^{-1}+\underline{\underline{T}}\,\underline{\underline{H}}\,\underline{\underline{T}}^{-1}. \tag{5.48}$$

Requiring therefore thet $\underline{\underline{H}}'$ satisfy the *hermiticity* condition (5.35) provides the equation defining the transformation matrix $\underline{\underline{T}}(\tau)$.

6. Invariants associated with ordinary differential equations.

We will limit the discussion to second order differential equation of the type (5.43), i. e.

$$\frac{d}{d\tau}[n_0(\tau)\dot{y}]+n_2(\tau)y=0, \tag{6.1}$$

which, as already noticed, rules the paraxial propagation of an optical beam through a non homogeneous medium or the paraxial motion of a charged particle along a transport channel.
As remarked in the previous section, Eq. (6.1) can be cast in the vector form (5.45) with

$$\underline{\underline{H}} = \begin{pmatrix} 0 & \dfrac{1}{n_0(\tau)} \\ -n_2(\tau) & 0 \end{pmatrix} \tag{6.2}$$

and consequently regarded as the Hamilton's equation of motion for the conjugate variables y_1 and y_2.

Let us address the problem of searching for a vector $\underline{s}$, which does not change with τ. In general, $\underline{s}$ can be obtained by acting with an appropriate matrix $\underline{\underline{T}}(\tau)$ on the 'instantaneous' vector $\underline{y}(\tau)$:

$$\underline{s} = \underline{\underline{T}}(\tau)\underline{y}(\tau), \tag{6.3}$$

whose derivative with respect to τ

$$\frac{d}{d\tau}\underline{s} = \left[\frac{d}{d\tau}\underline{\underline{T}} + \underline{\underline{T}}\,\underline{\underline{H}}\right]\underline{y} \tag{6.4}$$

provides the following equation of motion for $\underline{\underline{T}}$:

$$\frac{d}{d\tau}\underline{\underline{T}} = -\underline{\underline{T}}\,\underline{\underline{H}}, \tag{6.5a}$$

or equivalently

$$\frac{d}{d\tau}\underline{\underline{T}}^{-1} = \underline{\underline{H}}\,\underline{\underline{T}}^{-1}, \tag{6.5b}$$

which assures the τ-invariance of the vector $\underline{s}$. The initial condition to be associated with the above equations can be arbitrarily chosen. It is evident from (6.3) that assuming $\underline{\underline{T}}(0)$ to be the unit matrix:

$$\underline{\underline{T}}(0)=\underline{\underline{I}} \tag{6.6}$$

amounts to identify $\underline{\underline{T}}$ with the inverse of the displacement matrix: $\underline{\underline{T}}=\underline{\underline{U}}^{-1}$.

Consequently, the components of the invariant vector $\underline{s}$ are immediately understood as the initial values of the $\underline{y}$-components:

$$\underline{s}=\begin{pmatrix} y_1 & (0) \\ y_2 & (0) \end{pmatrix}. \tag{6.7}$$

The explicit expression of the $\underline{\underline{T}}$-matrix elements in terms of the (A,B,C,D) elements of the evolution matrix allows to specialize (6.7) as functional forms of the components of $\underline{y}$ according to

$$\underline{s}=\begin{pmatrix} D(\tau)y_1(\tau)-B(\tau)y_2(\tau) \\ -C(\tau)y_1(\tau)+A(\tau)y_2(\tau) \end{pmatrix}. \tag{6.8}$$

The above expression are easily recognized as the linear invariants, originally introduced by Lewis [30] in the anlysis of quantum states, driven by a time-dependent harmonic oscillator-type Hamiltonian. Later Dodonov and Man'ko [31] utilized the invariant (6.8) to introduce a new definition of coherent states, applied to both the dynamics af a damped

harmonic oscillator and the paraxial propagation of an e.m. wave through a non-homogeneous optical fiber.
In general, for an arbitrary choice of the initial values of $\underline{\underline{T}}$, the components of $\underline{s}$ can be expressed as a linear combination of the initial values $y_1(0)$ and $y_2(0)$, thus implying that any linear invariants can be expressed as a linear combination of the invariant (6.8).
It can be easily recognized that any function of the linear invariant (6.8) is itself an invariant. In particular, the quadratic form defined as

$$\epsilon(s_1, s_2) = s_1^2 + s_2^2 \tag{6.9}$$

is an integral of motion. According to (6.8), it can be expressed in terms of the instantaneous components of $\underline{y}$, i. e.

$$\epsilon = \gamma(\tau) y_1^2(\tau) + \beta(\tau) y_2^2(\tau) + 2\alpha(\tau) y_1(\tau) y_2(\tau), \tag{6.10}$$

where the functions α, β and γ are defined in terms of the $\underline{\underline{U}}$-matrix elements as

$$\alpha = -(AC + BD) \qquad \beta = A^2 + B^2 \qquad \gamma = D^2 + C^2. \tag{6.11}$$

In the phase plane (y_1, y_2) the form (6.10) represents an ellipse (*), whose axes make the angles

---

(*) It can be easily checked that the determinant relevant to the conic (6.10) is greater than zero, in particular we have $\beta\gamma - \alpha^2 = (AD - BC)^2 = 1$.

$$\phi = -\frac{1}{2}\tan^{-1}\left(2\frac{\alpha}{\beta-\gamma}\right) \qquad \theta = \frac{\pi}{2}+\phi \tag{6.12}$$

with the axis y_1 (see Fig. 2). The ellipse rotates in the phase-plane with a rate determined by the τ-dependence of the angle ϕ, correspondently the length of its axes changes with τ (*). However the area of the ellipse does not change with τ, being given by $\pi\epsilon$.

It is interesting to notice that the evolution of the ratio

$$q(\tau) = \frac{y_1(\tau)}{y_2(\tau)}, \tag{6.13}$$

which can be understood as the cotangent of the angle that the vector $\underline{y}$ makes with the y_1-axis at τ, is given by

$$q(\tau) = \frac{Aq_0 + B}{Cq_0 + D}. \tag{6.14}$$

(*) The maximum y_1 and y_2 excursions $y_{1\,max}$ and $y_{2\,max}$ are given by

$$y_{1\,max} = \sqrt{\epsilon\gamma} \qquad y_{2\,max} = \sqrt{\epsilon\beta}.$$

Correspondently, the coordinates of the intercepts with the axes are

$$y_{1i} = \sqrt{\frac{\epsilon}{\beta}} \qquad y_{2i} = \sqrt{\frac{\epsilon}{\gamma}}.$$

The above relation can be immediately recognized as the *ABCD* law, which rules the evolution of the complex curvature radius $q(z)$ of a gaussian beam propagating along an optical system described by the *ABCD* matrix.

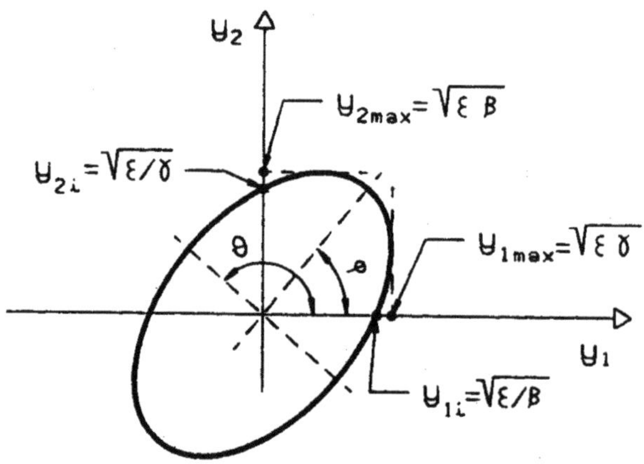

Fig. 2. Invariant (6.10) in the *phase* plane (y_1, y_2).

The quadratic form (6.10) resembles the Courant-Snyder invariant [32] introduced within the context of the analysis of the motion of a charged particle along a transport channel. Correspondently, the functions α, β and γ are easily related to the Twiss parameters, which play an important role in the design of a transport channel.

In addition, the expression (6.10) suggests to determine the conditions on the functions α, β and γ that a quadratic form of the type (6.10) be an integral of motion. It is well known that any bilinear form $I(\tau)$ in the components of $\underline{y}$ can be associated with a matrix $\underline{\underline{I}}(\tau)$ according to (*)

$$I(\tau)=\frac{1}{2}\underline{y}^T\underline{\underline{J}}\,\underline{\underline{I}}\,\underline{y}. \tag{6.15}$$

Consequently, we must specify the conditions on $\underline{\underline{I}}$ in order to assure the invariance of the corresponding quadratic form $I(\tau)$. According to (5.37), the derivative of (6.15) with respect to τ specializes as

$$\frac{d}{d\tau}I(\tau)=\frac{1}{2}\underline{y}^T\underline{\underline{J}}\left\{\frac{d}{d\tau}\underline{\underline{I}}-[\underline{\underline{H}},\underline{\underline{I}}]\right\}\underline{y}. \tag{6.16}$$

which, requiring the invariance of $I(\tau)$, provides the equation of motion for $\underline{\underline{I}}(\tau)$:

$$\frac{d}{d\tau}\underline{\underline{I}}(\tau)=[\underline{\underline{H}},\underline{\underline{I}}]. \tag{6.17}$$

Specializing therefore $\underline{\underline{I}}$ as (**)

(*) Conversely, the matrix $\underline{\underline{I}}$ associated with a quadratic form $I(\tau)$ can be recovered according to

$$\{y_1,I(\tau)\}=I_{11}(\tau)y_1+I_{12}(\tau)y_2 \qquad \{y_2,I(\tau)\}=I_{21}(\tau)y_1+I_{22}(\tau)y_2,$$

where {} denotes Poisson brackets.

(**) The trace of $\underline{\underline{I}}$ vanishes as a consequence of the condition $\underline{\underline{I}}^T\underline{\underline{J}}+\underline{\underline{J}}\,\underline{\underline{I}}=0$

$$\underline{\underline{I}}(\tau)=\begin{pmatrix}\alpha & \beta\\ -\gamma & -\alpha\end{pmatrix}, \tag{6.18}$$

which yields the quadratic form (6.10), provides the following set of differential equations for the functions α, β and γ:

$$\begin{aligned}\dot{\gamma}&=2n_2\alpha\\ \dot{\alpha}&=-\frac{\gamma}{n_0}+n_2\beta\\ \dot{\beta}&=-2\frac{\alpha}{n_0}.\end{aligned} \tag{6.19}$$

Furthermore, the relation (6.15) provides a useful method to get the link between the values of the functions α, β and γ at different τ. In fact, the condition for the invariance of $I(\tau)$ explicitly writes

$$\underline{y}_1^T\underline{\underline{J}}\,\underline{\underline{I}}(\tau_1)\underline{y}_1=\underline{y}_2^T\underline{\underline{J}}\,\underline{\underline{I}}(\tau_2)\underline{y}_2 \tag{6.20}$$

with $\underline{y}_1$ and $\underline{y}_2$ denoting the vector $\underline{y}$ at τ_1 and τ_2 respectively. Expressing therefore $\underline{y}_1$ in terms of $\underline{y}_2$ as

$$\underline{y}_1=\underline{\underline{U}}^{-1}(\tau_2,\tau_1)\underline{y}_2 \tag{6.21}$$

and exploiting the relation (5.33) we get

$$\underline{\underline{I}}(\tau_2)=\underline{\underline{U}}(\tau_2,\tau_1)\underline{\underline{I}}(\tau_1)\underline{\underline{U}}^{-1}(\tau_2,\tau_1), \tag{6.22}$$

which can be easily recognized as a similarity transformation. Consequently, we can construct a transport matrix, whose elements are related to the (A,B,C,D) elements of the displacements matrix as

$$\begin{pmatrix} \alpha_2 \\ \beta_2 \\ \gamma_2 \end{pmatrix} = \begin{pmatrix} AD+BC & AC & BD \\ 2AB & A^2 & B^2 \\ 2DC & C^2 & D^2 \end{pmatrix}. \tag{6.23}$$

The relation (6.22) states the invariance of the determinant of $\underline{\underline{I}}$. In fact, since $det\underline{\underline{U}}$ is equal to unit, we get that

$$det\underline{\underline{I}}(\tau) = constant, \tag{6.24}$$

the specific value being determined by the initial values of α, β and γ.
However, the further condition (5.33) is usually required to be satisfied by the matrix $\underline{\underline{I}}$, thus implying that

$$\underline{\underline{I}}^2 = -\underline{\underline{1}}, \tag{6.25}$$

which allows to specialize the determinant of $\underline{\underline{I}}$ as

$$det\underline{\underline{I}}(\tau) = \pm 1. \tag{6.26}$$

In order to solve the ambiguity on the sign of the determinant of $\underline{\underline{I}}$, let us comment on the meaning of the Twiss parameters α, β and γ.

According to (5.38), the functional forms associated with the matrices (4.13), which realize the matrix representation of the SU(1,1) generators (*), are easily inferred as (‡)

$$K_3(y_1,y_2,\tau)=-\frac{1}{2}y_1y_2,\quad K_+(y_1y_2,\tau)=\frac{1}{2}y_1^2,\quad K_-(y_1,y_2,\tau)=\frac{1}{2}y_2^2, \tag{6.27}$$

which within the context of a multiparticle dynamics are easily associated with the quadratic moments of the particle distribution.
The evolution of the above functions is ruled by the equations

$$\begin{aligned}&\frac{d}{d\tau}K_+=-\frac{2}{n_0}K_3\\&\frac{d}{d\tau}K_3=n_2K_+-\frac{1}{n_0}K_-,\\&\frac{d}{d\tau}K_-=2n_2K_3\end{aligned} \tag{6.28}$$

which once compared with (6.19) provides the following correspondences with the Twiss parameters

$$K_3 \Leftrightarrow \alpha \quad K_+ \Leftrightarrow \beta \quad K_- \Leftrightarrow \gamma \tag{6.29}$$

(*) Let us recall that for the two-dimensional case, we are considering, the algebra sp(2,R) is isomorphic to the SU(1,1) algebra.

(‡) It is worth stressing the structural similarity between the function (6.27) and the representation of the operators $\hat{K}_3$, $\hat{K}_\pm$ as given in (3.5) or (4.11).

Consequently, according to (6.27) it is immediate to recognize the following relation

$$det\underline{\underline{I}} \Leftrightarrow <y_1^2><y_2^2>-<y_1 y_2>^2, \tag{6.30}$$

which according to the Schwarz inequality allows to choose the positive sign in (6.26).

On the other hand, the above-recognized correspondences suggest the possibility of relating the Twiss parameters α, β and γ with the components of a Bloch-type vector, usually introduced within the context of the analysis of the time-evolution of quantum states driven by SU(2) or SU(1,1)-type Hamiltonian operators [22].

Introducing indeed the functions

$$K_1(\tau)=\frac{1}{2}(K_+ + K_-) \qquad K_2(\tau)=\frac{1}{2}(K_+ - K_-), \tag{6.31}$$

it is immediate to get from (6.28), or equivalently from (6.19) with the aid of the correspondences

$$K_1 \Leftrightarrow \frac{1}{2}(\beta+\gamma) \qquad K_2 \Leftrightarrow \frac{1}{2}(\beta-\gamma), \tag{6.32}$$

the following Bloch-type equation for the vector $\underline{K}=(K_1, K_2, K_3)$:

$$\frac{d}{d\tau}K_j=\tilde{\epsilon}_{lmj}Q_l K_m, \tag{6.33}$$

with $\tilde{\epsilon}_{lmj}$ being the *Ricci tensor* relevant to the Poisson brackets $\{K_i, K_j\}$ and Q_l denoting the components of the vector

$$\underline{Q} = \left(\frac{1}{n_0} + n_2, \frac{1}{n_0} - n_2, 0\right). \tag{6.34}$$

Equation (6.32) can be visualized as a pseudoprecession motion of the vector $\underline{K}$ around the instantaneous direction of $\underline{Q}$ within a non-euclidean space, characterized by the metric matrix

$$\underline{\underline{M}} = \begin{pmatrix} 1 & 0 & 0 \\ 0 & -1 & 0 \\ 0 & 0 & -1 \end{pmatrix}. \tag{6.35}$$

Accordingly, it can be easily verified that the norm of $\underline{K}$ defined

$$K^2 = \underline{K}^T \underline{\underline{M}}\,\underline{K} = K_1^2 - K_2^2 - K_3^2 \tag{6.36}$$

does not change with time, thus providing the law of conservation: $K^2 = constant$

7. Motion of a charged particle in a circular accelerator.

The particles in a circular magnetic accelerator, such as a sinchrotron, ciclotron or betatron, are confined near their equilibrium orbit by magnetic focusing forces [33]. The magnets are usually disposed along a roughly circular closed orbit, which is usually referred as *ideal* (or *equilibrium*) orbit (*). The guide magnetic field must be designed in order to assure that the motion of the particle near the equilibrium orbit is stable in the sense that if a particle, whose momentum is appropriate to the given ideal orbit, is started with a small initial displacement and a small initial angle from the equilibrium orbit, it will remain near the ideal orbit for all time.

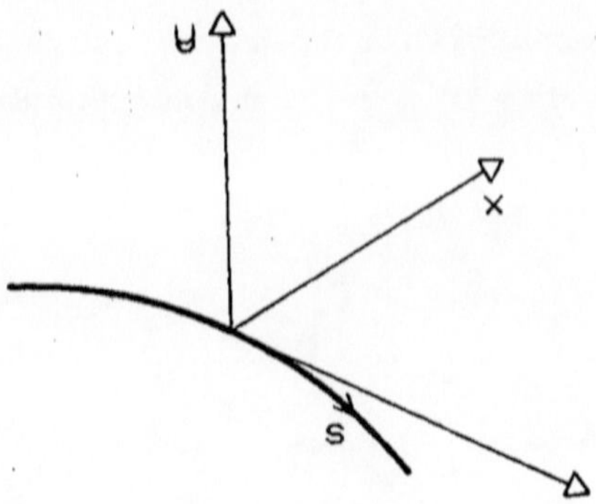

Fig. 3. Coordinate system for the motion in a circular accelerator.

The motion of a particle in a circular accelerator is conveniently described in terms of coordinates related to the ideal orbit. The instantaneous position of the particle will be specified indeed by the circular coordinates (s,x,y), where s is the distance along the ideal orbit measured from some fixed point to the point closed to the particle, x and y are

(*) Within the present context, the following terminology is usually adopted: a *trajectory* is any particle path, while an *orbit* is a particular trajectory, which repeats itself on successive revolutions.

the horizontal and vertical component of the particle's displacement from the equilibrium orbit. The positive sense of s is usually taken in the sense of the particle's motion, of x in the outward direction and y in the upward direction (see Fig. 3).

The coordinates x and y are assumed *small* with respect to the local curvature radius of the trajectory. As a consequence, in evaluating the forces acting on the particle, only linear terms in x and y need be retained. Within the context of this linear approximation, the equations of motion for a particle with no energy error with respect to the nominal value can be expressed in terms of s as the independent coordinate in the general form [34]

$$\begin{aligned} &\ddot{x} + K_x(s)x = Qy - 2L\dot{y} \\ &\ddot{y} + K_y(s)y = -Qx + 2L\dot{x}, \end{aligned} \tag{7.1}$$

where the dot denotes differentiation with respect to the longitudinal coordinate s. The terms on the right hand of Eq. (7.1) account for the coupling between the radial and vertical motion as due to the skew quadrupoles (term proportional to Q) and to the solenoids (term proportional to L). The quantities K_x and K_y, whose possible dependence on the longitudinal coordinate s has been explicitly pointed out, include the interaction of the particle with both the normal quadrupoles, which provide the guide magnetic field, and the solenoids. Introducing the vector

$$\underline{r} \equiv (x, p_x, y, p_y), \tag{7.2}$$

where (x, p_x) and (y, p_y) are conjugate variables, we can write Eq. (7.1) in the vector form

$$\frac{d}{ds}\underline{r}(s) = \underline{\underline{H}}(s)\underline{r}(s). \tag{7.3}$$

The 4×4 matrix $\underline{\underline{H}}$, explicitly given by

$$\underline{\underline{H}} = \begin{pmatrix} 0 & 1 & L & 0 \\ -k_x(s) & 0 & -Q & L \\ -L & 0 & 0 & 1 \\ -Q & -L & -k_y(s) & 0 \end{pmatrix}, \tag{7.4}$$

satisfies the relation (5.35) with the symplectic matrix $\underline{\underline{J}}$ being

$$\underline{\underline{J}} = \begin{pmatrix} 0 & -1 & 0 & 0 \\ 1 & 0 & 0 & 0 \\ 0 & 0 & 0 & -1 \\ 0 & 0 & 1 & 0 \end{pmatrix}. \tag{7.5}$$

The Hamiltonian associated with the dynamics described by Eq. (7.1) is then easily obtained according to (5.38) as

$$H(x,p_x,y,p_y) = \frac{1}{2}(p_x^2 + k_x x^2) + \frac{1}{2}(p_y^2 + k_y y^2) + Qxy + L(yp_x - xp_y), \tag{7.6}$$

where as in (7.4) $k_x(s)$ and $k_y(s)$ include only the effect of the guide magnetic field. They are related to the field gradient index and to the local curvature radius of the ideal orbit. For storage rings, k_x and k_y are approximately equal and have opposite signs, the guide magnetic field, therefore, is focusing in one direction and defocusing in the other one.

Let us consider first the case where the coupling terms in (7.6) are negligible. As a consequence, the matrix $\underline{\underline{H}}$ assumes a block-structure, thus allowing to analyse separately the motion along the horizontal and vertical directions. We must deal therefore with the Hill's equation

$$\frac{d^2}{ds^2}y(s)+k(s)y(s)=0, \tag{7.7}$$

where the restoring force is periodic:

$$k(s)=k(s+l), \tag{7.8}$$

the period l being determined by the length of the 'unit' cells, which compose the circular machine, we are considering. The vector version of Eq. (7.6) is immediately written down as

$$\frac{d}{ds}\underline{y}(s)=\begin{pmatrix} 0 & 1 \\ -k(s) & 0 \end{pmatrix}\underline{y}(s), \tag{7.9}$$

the components of the vector $\underline{y}$ being the local height y and the slope $\dot{y}=\frac{d}{ds}y$ with respect to the reference orbit of the projection of the particle trajectory into the horizontal and vertical plane. This description is formally similar to that adopted in ray-optics, the ray-parameters being, as noticed in Sec. 4, the height and the reduced slope of the ray with respect to the optical axis. Equation (7.7), which as already remarked is relevant to a linear approximation, resembles indeed the paraxial ray-equation, relevant to a non-homogeneous medium exhibiting a unit refractive index on the optical axis ($n_0(z)=1$ according to the notation of Sec. 4 or 5).

Applying the formalism developed in the previous sections, we can introduce the displacement matrix, usually reported within the present context as the *transfer* matrix $\underline{\underline{M}}$, in terms of which we can formally write the solution to (7.9) as

$$\underline{y}(s)=\underline{\underline{M}}(s,s_0)\underline{y}(s_0). \tag{7.10}$$

According to the discussion carried out in Sec. 5, the elements of $\underline{\underline{M}}$ are the cos-type $C(s,s_0)$ and sin-type $S(s,s_0)$ solutions to Eq. (7.7):

$$\underline{\underline{M}}(s,s_0)=\begin{pmatrix} C(s,s_0) & S(s,s_0) \\ \dot{C}(s,s_0) & \dot{S}(s,s_0) \end{pmatrix}, \tag{7.11}$$

where the notation usually adopted within the context of accelerator physics has been introduced.
The Wei-Norman theorem allows to express $\underline{\underline{M}}$ as a product of matrices

$$\underline{\underline{M}}(s,s_0)=\exp\{2h\underline{\underline{K}}_3\}\exp\{g\underline{\underline{K}}_+\}\exp\{f\underline{\underline{K}}_-\}, \tag{7.12}$$

$\underline{\underline{K}}_3$, $\underline{\underline{K}}_\pm$ being the generators of the SU(1,1) algebra (or equivalently $sp(2,\mathbb{R})$), which, as already noticed, is naturally associated with the problem under study. As in ray-optics (see Sec. 4), each matrix entering the above expression describes the effect of specific elements in the transport channel, as it will be illustrated in the following.

Just to make an interesting comparison with the ray-matrix (4.23) relevant to the paraxial propagation through a medium with n_2 and n_0 constant (in particular, (4.23) refers to a focusing profile of the refractive index), let us write down the transfer matrix for a segment in which k is constant:

$$\underline{\underline{M}}(s,s_0)=\begin{pmatrix} \cos\phi & -\frac{1}{\sqrt{k}}\sin\phi \\ \sqrt{k}\sin\phi & \cos\phi \end{pmatrix}, \tag{7.13}$$

with $\phi=\sqrt{k}(s-s_0)$ and k being greater or less than zero for focusing or defocusing segments respectively.

Let us turn now to consider the invariant forms associated with Eq. (7.7). As noticed in Sec. 6, the quadratic expression (6.10) in the phase plane variables $(y,\dot{y})$ is invariant. In the language of accelerator physics one usually deals with the beam-ellipse matrix $\underline{\underline{\Sigma}}$ (*) which is related to the matrix $\underline{\underline{I}}$ introduced in Sec. 6 according to

$$\underline{\underline{\Sigma}}(s)=-\underline{\underline{I}}\,\underline{\underline{J}}=\begin{pmatrix} \beta(s) & -\alpha(s) \\ -\alpha(s) & \gamma(s) \end{pmatrix}, \tag{7.14}$$

(*) According to the Liouville's theorem, a given charged-particle beam can be associated with an invariant area in the phase plane $(y,\dot{y})$, which is usually reported as the emittance ϵ of the particle beam. When the beam-ellipse is concentric and similar to the machine-ellipse, the beam is said to be matched to the machine. In this instance, the beam reappears an successive turns as the same ellipse, as a consequence of the periodicity of the motion along the circular machine.

with the specific requirement on the determinant of $\underline{\underline{\Sigma}}$ that

$$det\underline{\underline{\Sigma}} = \beta\gamma - \alpha^2 = 1. \tag{7.15}$$

Within this concern, the ellipse associated with the matrix $\underline{\underline{\Sigma}}$ is obtained as

$$\epsilon = \underline{y}^T \underline{\underline{\Sigma}}^{-1} \underline{y} = \gamma(s) y^2 + \beta(s) \dot{y}^2 + 2\alpha(s) y \dot{y}, \tag{7.16}$$

wher ϵ, which specifies the area of the ellipse, is usually referred as the emittance of the beam.
As the motion progresses with transfer matrix $\underline{\underline{M}}$, the beam ellipse transforms like

$$\underline{\underline{\Sigma}}(s) = \underline{\underline{M}}(s, s_0) \underline{\underline{\Sigma}}(s_0) \underline{\underline{M}}(s, s_0). \tag{7.17}$$

In this instance, the equations (6.19) defining the evolution of the Twiss parameters α, β and γ specialize into

$$\begin{aligned} \dot{\beta} &= 2\alpha \\ \dot{\alpha} &= k\beta - \gamma, \\ \dot{\gamma} &= 2k\alpha \end{aligned} \tag{7.18}$$

which can be appropriately combined to get the following third order equation for β:

$$\dddot{\beta} + 4k\dot{\beta} + 2\dot{k}\beta = 0. \tag{7.19}$$

However the relation (7.15) provides the following

$$\alpha = \frac{\dot{\beta}}{2} \qquad \gamma = \frac{1+\alpha^2}{\beta}. \tag{7.20}$$

Let us consider now the matrices entering the expression of $\underline{\underline{M}}$ as given in (7.12) with refernce to the effect on the Twiss parameters and correspondently on the beam-ellipse. It is interesting to notice that each of the matrices (7.12) leaves one of the Twiss parameters unchanged.
According to the discussion carried out in Sec. 4, within the context of ray-optics, the matrix

$$e^{f\underline{\underline{K}}_+} = \begin{pmatrix} 1 & f \\ 0 & 1 \end{pmatrix} \tag{7.21}$$

describes the effect of a drift space or field-free region. Accordingly, the transfer matrix for the Twiss parameters as given in (6.24) specializes as

$$\begin{pmatrix} \alpha_2 \\ \beta_2 \\ \gamma_2 \end{pmatrix} = \begin{pmatrix} 1 & 0 & f \\ 2f & 1 & f^2 \\ 0 & 0 & 1 \end{pmatrix} \begin{pmatrix} \alpha_1 \\ \beta_1 \\ \gamma_1 \end{pmatrix}, \tag{7.22}$$

which clearly displays that during the motion through a field free region the parameter γ remains unchanged.
Consequently, as regards to the beam-ellipse, the maximum y excursion and the coordinate of the intercept with the $\dot{y}$ axis do not change, as it can be seen in Fig. 4, where the evolution of the beam-ellipse is sketched.
On the other hand, the transfer matrix

$$e^{g\underline{K}_-}=\begin{pmatrix}1 & 0\\ -g & 1\end{pmatrix} \qquad (7.23)$$

describes the effect of a thin lens with focus $F=-\frac{1}{g}$.

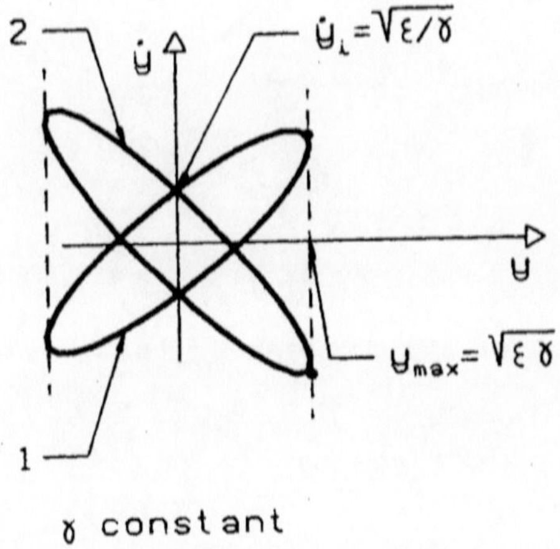

Fig. 4. Beam-ellipse evolution along the drift-space.

The Twiss parameters transform according to

$$\begin{pmatrix}\alpha_2\\ \beta_2\\ \gamma_2\end{pmatrix}=\begin{pmatrix}1 & -g & 0\\ 0 & 1 & 0\\ -2g & g^2 & 1\end{pmatrix}\begin{pmatrix}\alpha_1\\ \beta_1\\ \gamma_1\end{pmatrix}, \qquad (7.24)$$

thus pointing out that the *propagation through a thin lens* does not affect the value of β. The transformation of the beam-ellipse is illustrated in Fig. 5. Practically, a quadrupole behaves in each phase-plane $(x,\dot{x})$ and $(y,\dot{y})$ like a thin lens of opposite sign, that is focusing in one direction and defocusing in the other one.

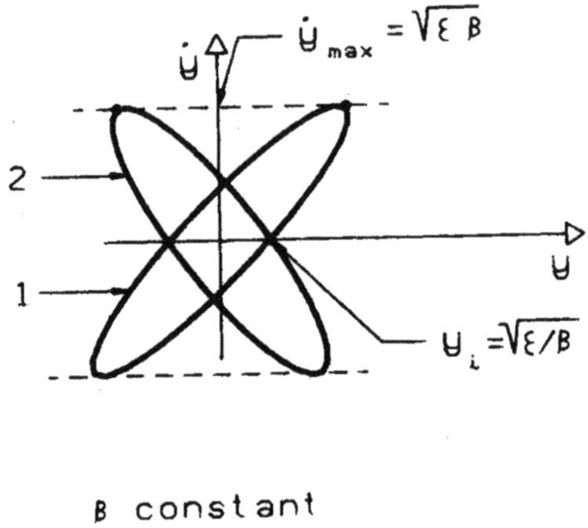

Fig. 5. Effect of a thin lens on the beam-ellipse.

Finally, let us consider the matrix

$$e^{2h\underline{K}_3} = \begin{pmatrix} M & 0 \\ 0 & \dfrac{1}{M} \end{pmatrix}, \qquad M = e^{-h}, \tag{7.25}$$

whose effect can be accounted for by an appropriate combination of drift-space and thin-lens quadrupoles according to the scheme of Fig. 1. It can be easily inferred from the transfer matrix for the Twiss parameters:

$$\begin{pmatrix}\alpha_2\\ \beta_2\\ \gamma_2\end{pmatrix} = \begin{pmatrix}1 & 0 & 0\\ 0 & M^2 & 0\\ 0 & 0 & \dfrac{1}{M^2}\end{pmatrix}\begin{pmatrix}\alpha_1\\ \beta_1\\ \gamma_1\end{pmatrix} \tag{7.26}$$

that a telescopic system, described by the matrix (7.25), does not change the value of α. The effect on the beam-ellipse is shown in Fig. 6.

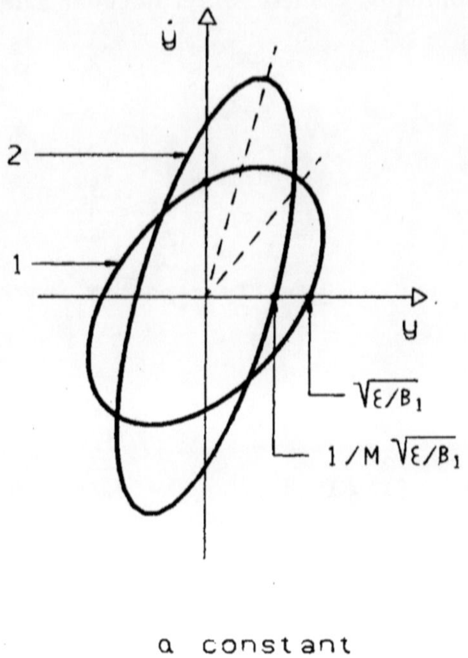

Fig. 6. Beam-ellipse after the propagation through a telescopic system.

Let us turn now to the explicit expression of the solution to Eq. (7.7) or (7.9), which according to the Floquet theorem can be written in the form of a 'quasi-harmonic' oscillation

$$y(s)=\sqrt{\epsilon\beta(s)}\cos(\psi(s)+\phi), \tag{7.27}$$

with varying amplitude $\sqrt{\epsilon\beta(s)}$ and varying instantaneous wavelength $\lambda=2\pi\left(\frac{d}{ds}\psi\right)^{-1}$.
The angle ϕ is defined by the initial conditions and $\psi(s)$, referred as the *machine phase shift* between the points s_0 and s is given by

$$\psi(s)=\int_{s_0}^{s}\frac{1}{\beta(s')}ds'. \tag{7.28}$$

Correspondently, $\dot{y}(s)$ is usually reported in the form

$$\dot{y}(s)=\sqrt{\epsilon\gamma(s)}\cos(\chi(s)+\phi), \tag{7.29}$$

with the function $\chi(s)$ satisfying the relation

$$\tan^{-1}(\psi-\chi)=\frac{1}{\alpha}. \tag{7.30}$$

The average over the initial angle ϕ provides the moments of the particle distribution. In particular, it is readily verified that the quadratic moments are related to the emittance ϵ and to the β and γ Twiss parameters according to

$$\sigma(s) = \langle y^2 \rangle^{\frac{1}{2}} = \frac{1}{2}\sqrt{\epsilon\beta(s)}$$

$$\dot{\sigma}(s) = \langle \dot{y}^2 \rangle^{\frac{1}{2}} = \frac{1}{2}\sqrt{\epsilon\gamma(s)}, \tag{7.31}$$

σ and $\dot{\sigma}$ giving the r.m.s. width and divergence of the particle beam.

It is interesting to notice that a kind of uncertainity principle can be stated for σ and $\dot{\sigma}$. The product

$$\sigma\dot{\sigma} = \frac{1}{4}\epsilon\sqrt{\beta\gamma} = \frac{1}{4}\epsilon\sqrt{1+\alpha^2} \tag{7.32}$$

is indeed always equal to or greater than the emittance ϵ. For $\alpha = 0$, we get a minimum uncertainity beam. Similarly, a particle distribution can exhibit squeezing in one of the two conjugate variables y or $\dot{y}$ according to whether

$$\sigma^2 < \frac{1}{4}\epsilon\sqrt{1+\alpha^2} \iff \beta < 1+\alpha^2 \qquad y-\textit{squeezing}$$

$$\dot{\sigma}^2 < \frac{1}{4}\epsilon\sqrt{1+\alpha^2} \iff \gamma < 1+\alpha^2 \qquad \dot{y}-\textit{squeezing}. \tag{7.33}$$

Within this concern, a coherent beam could be understood as a symmetric minimum uncertainity beam, characterized then by the relations

$$\sigma\dot{\sigma} = \frac{\epsilon}{4} \qquad \sigma^2 = \dot{\sigma}^2 = \frac{\epsilon}{4}. \tag{7.34}$$

Finally, it is easy to get the equation governing the evolution of σ as (*)

$$\ddot{\sigma} + k\sigma - \frac{\epsilon^2}{\sigma^3} = 0, \tag{7.35}$$

which is usually reported as the *Kapchinskij-Vladimirinskij* equation [35].
Before concluding this brief discussion of the uncoupled transverse motion of a charged particle along a circular accelerator, let us point out the possibility of expressing the emittance ϵ in terms of the quadratic moments as

$$\epsilon = \sqrt{<y^2><\dot{y}^2> - <y\dot{y}>^2}. \tag{7.36}$$

In this instance, it is worth noticing that ϵ migth be obtained from the eigenvalue of the *covariance* matrix [36,37], defined as

$$\begin{pmatrix} -<y\dot{y}> & <y^2> \\ -<\dot{y}^2> & <y\dot{y}> \end{pmatrix}, \tag{7.37}$$

which, according to the correspondences stated in (6.29), is related to the Twiss matrix $\underline{\underline{I}}$ or $\underline{\underline{\Sigma}}$. Finally, let us notice

(*) For a discussion of the envelope equation (7.35) within the context of the propagation through an optical fiber, see ref. [18].

that the norm of the Bloch vector introduced in Sec. 6, whose components are defined in terms of the quadratic moments as indicated in (6.27), is nothing but $\frac{\epsilon^2}{4}$.

So far we have discussed the decoupled x, y motion; if the solenoidal or skew quadrupoles are active, the extension of the above formalism to two degree of freedom is in order. This extension is conceptually straightforward and presents some computational difficulties only, since as already noticed $4x4$ matrices should be employed. The equations of motion for the coupled system are provided by Eqs. (7.3) for the four vector (7.2). The associated Twiss matrix can be now written in the form

$$\underline{\underline{I}} = \begin{pmatrix} \underline{\underline{I}}_x & \underline{\underline{I}}_{x,y} \\ \underline{\underline{\tilde{I}}}_{x,y} & \underline{\underline{I}}_y \end{pmatrix}, \tag{7.38}$$

where the entries of $\underline{\underline{I}}$ are $2x2$ matrices, whose equations of motion can be inferred from (6.17) as

$$\begin{aligned}
\frac{d}{ds}\underline{\underline{I}}_x &= \left[\underline{\underline{H}}_x, \underline{\underline{I}}_x\right] + \left\{\underline{\underline{H}}_{x,y}\underline{\underline{\tilde{I}}}_{x,y} - \underline{\underline{I}}_{x,y}\underline{\underline{\tilde{H}}}_{x,y}\right\} \\
\frac{d}{ds}\underline{\underline{I}}_{x,y} &= \underline{\underline{H}}_x\underline{\underline{I}}_{x,y} + \underline{\underline{H}}_{x,y}\underline{\underline{I}}_y - \underline{\underline{I}}_x\underline{\underline{H}}_{x,y} - \underline{\underline{I}}_{x,y}\underline{\underline{H}}_y, \\
\frac{d}{ds}\underline{\underline{I}}_y &= \left[\underline{\underline{H}}_y, \underline{\underline{I}}_y\right] + \left\{\underline{\underline{\tilde{H}}}_{x,y}\underline{\underline{I}}_{x,y} - \underline{\underline{\tilde{I}}}_{x,y}\underline{\underline{H}}_{x,y}\right\}.
\end{aligned} \tag{7.39}$$

The above equations are the direct generalization of those relevant to the Twiss coefficients and can be conveniently exploited to characterize the evolution of the particle beam. In particular, as already noticed, the off-diagonal elements of the matrix $\underline{\underline{I}}$ are related to the beam size and divergence. However, for a more detailed discussion the interested reader is addressed to ref. [38].

In the present paper we have shown how a purely algebraic formalism can provide the natural tool of analysis to treat a variety of apparently unrelated problems in optics. We have essentially recovered well known results within the framework of a unifying method but, as elsewhere shown ref. [39], the present method can be conveniently applied when non linear effects arise. We must however underline that group theoretic methods are powerful because of their compactness and universality, but otherwise the same results could be recovered with much more effort using conventional tools.

Acknowledgements

During the compilation of the paper we enjoyed many helpful discussions with our collegues at the ENEA-Frascati Center. In particular, it is a pleasure to recognize the continuous help and criticisms by drs. C. Mari, M. Richetta and C. Ronsivalle. Furthermore the editorial help of Mr. E. Campana and D. Vizzarri is also greatefully acknowledged.

References

[1] E. P. Wigner, Phys. Rev. 51 , 106 (1937)

[2] P. Franzini and L. Radicati, Phys. Lett. 6 , 322 (1963)

[3] J. P. Elliott, Proc. R. Soc. London Ser. A245 ,128 (1958)

[4] M. Gell'mann and Y. Ne'eman, *The eightfold way*, (Benjamin, New York, N. Y., 1967)

[5] J. N. Elgin, Phys Lett. 80A , 140 (1980)

[6] F. T. Hioe and J. H. Eberly, Phys. Resv. Lett. 47 , 838 (1981)

[7] H. Goldstein, *Classical Mechanics* (Addison Wesley Reading, Mass, 1959)

[8] G. Dattoli and A. Torre, to be published in JOSA B

[9] A. Vander Lugt, Proc. IEEE 54 , 1055 (1966)

[10] M. Nazarathy and J. Shamir, J. Opt. Soc. Am. 70 , 2 (1980); H. Bacry and M. Cadilhac, Phys. Rev. A23 , 2533 (1981); G. Dattoli,S. Solimeno and A. Torre, Phys. Rev. A35 , 1668 (1987)

[11] N. Vilenkin, 'Special functions and the theory of group representation', Amer. Math. Soc., Providence R. I. (1968)

[12] S. Lie, *Differentialgleichungen*, (Teubner, Leipzig, 1891)

[13] E. Schrodinger, Proc. Roy. Irish Acad. A46 ,9 (1940)

[14] W. Miller Jr, 'On Lie algebras and some special functions of mathematical physics', Amer. Math. Soc. Providence, R. I. (1964)

[15] S. Steinberg, J. Differential Eq. 26 , 404 (1977)

[16] G. Dattoli, M. Richetta, G. Schettini and A. Torre, to be published in J. Math. Phys.

[17] G. Dattoli, P. Di Lazzaro and A. Torre, to be published in Nuovo Cim. B

[18] G. Dattoli, V. Loreto, C. Mari, M. Richetta and A. Torre, *Biunitary transformations & ordinary differential equations*, Part I, II, to be pubished in Nuovo Cim. B

[19] W. Magnus, Commun. Pure Appl. Math. 7 , 649 (1954)

[20] same as [18], Part III

[21] R. M. Wilcox, J. Math. Phys. 4 , 962 (1967)
[22] G. Dattoli, J. C. Gallardo and A. Torre, La rivista del Nuovo Cim., 11 ,1 (1988)
[23] J. Wei and E. Norman, J. Math. Phys. 4 , 575 (1963)
[24] G. Dattoli, M. Richetta and A. Torre, Phys. Rev. A34 , 2466 (1986)
[25] G. Dattoli, M. Richetta and A. Torre, J. Math. Phys.,29 , 2586 (1988)
[26] G. Dattoli, R. Mignani and A. Torre, to be published in J. Math. Phys.
[27] D. Gloge and D. Marcuse, J. Opt. Soc. Am. 59 , 1629 (1969)
[28] A. Yariv, Quantum Electronics, (J. Wiley, New York, N. Y. 1975)
[29] E. L. Ince, *Ordinary Differential Equations*, Dover Publications, Inc. N. Y. (1956)
[30] H. R. Lewis, J. Math. Phys. 9 , 1976 (1968); H. R. Lewis and W. B. Riesenfeld, J. Math. Phys. 10 , 1458 (1969)
[31] V. V. Dodonov and V. I. Man'ko, Phys. Rev. A20 , 550 (1979); V. I. Man'ko, in 'Lie methods in Optics', Proceedings of the LEON conference 1985, edited by J. Sauchez Mondragon and K. B. Wolf (Springer-Verlag, Berlin, 1986).
[32] E. D. Courant and S. H. Snyder, Ann. Phys. 3 , 1 (1958)
[33] M. Sands, *The Physics of Electron Storage Rings*, SLAC-121-UC-28 (ACC)
[34] L. C. Teng, in *Frontiers of Particle Beams: Observation, Diagnosis & Correction*, Eds. M. Month and S. Turner, (Springer-Verlag, Berlin 1989)
[35] J. D. Lawson, *The physics of charged-particle beams*, (Clarendon Press, Oxford, 1978)
[36] F. J. Narcowich, J. Math. Phys. 31 , 354 (1990)
[37] F. Neri and G. Rangarajan, Phys. Rev. Lett. 64 , 1073 (1990)
[38] G. Dattoli, C. Mari, M. Richetta, C. Ronsivalle and A.

Torre, 'Classical and Quantum Optics from the point of view of Accelerator Physics', to be published in the Proceedings of the 'Workshop on nonlinear problems in future particle accelerators, Capri, 19-25 April 1990.
[39] G. Dattoli, C. Mari and A. Torre, 'A short note on the use of the evolution operator in classical mechanics', to be published in Nuovo Cim. B

Some Remarks about Path Integral Approach to Anomalies in Quantum Field Theory

Guido Cognola, Luciano Vanzo and Sergio Zerbini
Dipartimento di Fisica - Università di Trento - 38050 POVO - Italia
Istituto Nazionale di Fisica Nucleare - Gruppo Collegato di Trento

Abstract: *Some methods for calculating the expression of the anomaly starting from path integral are discussed in some detail. The link with zeta function and heat kernel expansion is pointed out. As applications, some known anomalies are explicitly evaluated.*

1 Introduction

The relevance of symmetry as a guiding principle in physics is well known. For example, if a classical physical system is described by an action which is invariant with respect to a rigid (or global) group of transformations, the Noether theorem leads to conservation laws. Usually this classical symmetry survives during the quantization process. Two examples are Poincarè and abelian gauge symmetries. However, when one is dealing with quantum field theory applied to the elementary particles physics, soft symmetry or spontaneous symmetry breaking are considered (Goldstone theorem, Nambu-Jona Lasinio and Higgs mechanism-see for example ref.[1]). As a result, a large number of phenomenas in the low energy processes as well as unified models for weak electromagnetic and strong interaction are reasonably described. In some cases a classical preexistent symmetry does not survive quantization in a manner which is completely different from the symmetry breaking mentioned above. The reason for that seems to stem from the following facts.

The perturbative quantum field theory (QFT) suffers from the presence of ultraviolet (maybe also infrared) divergences, mainly due to the short distance singularity of the free propagators of the fields involved. As a consequence, one needs a regularization and a renormalization procedure. Usually, the starting point is a classical theory, described by an action which can be invariant with respect to a global (maybe also local) gauge symmetry. It is convenient to choose a regularization which preserves this classical symmetry, but it may happen that another classical symmetry is broken at quantum level by the regularization scheme chosen. If it is the case, we say that an anomaly is present.

In 1969, Adler[2] and Bell and Jackiw[3] discovered the abelian chiral anomaly in connection with neutral pion decay, and this was the first anomaly which apperared in QFT. Since then, there has been a large amount of investigations about anomalies and on their role in the problem of internal consistency of gauge theories[4,5]. The current anomalies may be roughly divided in two classes: those associated with the currents coupled to the gauge fields and those associated with currents which do not couple to gauge fields (chiral anomaly). Since in the first case, the presence of anomalies seems to destroy the validity of Ward identities, which in turn raises problems associated with renormalizsability and unitarity, a general attitude has been to try to cancel them. As a result, a number of physically significant predictions comes out, among them the same number of quarks and leptons, the $N_c = 3$ prediction in QCD, t'Hooft instanton calculation[6] and so on. It is also whorth mentioning the superstring theory proposal where the cancellation of anomalies is achieved only for very limited number of gauge groups ($SO(32)$ and $E8 \times E8$)[7]. More recently, a general geometric setting has been used in the anomaly problem and cohomological tecniques have been introduced[8,9].

In section 2 we discuss two path integral approaches to anomalies, which have some interesting features and allows a non perturbative treatment of the problem. The link between anomalies, zeta function and spectral coefficients is pointed out. In section 3 we summarize an easy method for computing the heat coefficients explicitly. In section 4 we consider some significative examples of anomalies, that is the chiral-axial non abelian and conformal ones. We end the paper with some comments in section 5. For a more detailed treatment of such problems we refer the reader to the literature.

2 Anomalies

For sake of simplicity, we shall confine ourselves to the external field case, namely only matter fields ϕ, bosons or fermions, are quantized. To begin with, let us denote by

$$W[\mathcal{V}] = -\ln Z[\mathcal{V}] = -\ln \int \mathcal{D}(\phi)\, e^{-S[\phi,\mathcal{V}]} \tag{1}$$

the connected Euclidean Green functions generating functional, in which $\mathcal{D}(\phi)$ is the functional measure, whish we assume to be diffeomorphism invariant [10], $\mathcal{V}$ the set of external fields (for example a metric tensor, an axial vector field, and so on) and $S[\phi]$ the classical action given by

$$S[\phi] = \frac{1}{2} \int dvol\, \phi H \phi \tag{2}$$

where H is at least a second order self-adjoint differential operator, which contains the geometrical external fields. It can happen that the action is invariant with respect to a global (may be also local) symmetry; as a consequence, there exists a classical quantity A which is vanishing (for example the divergence of a current or the trace of a tensor). As we have seen in the introduction, quantum effects may lead to a non vanishing mean value of the corresponding quantum operator $\hat{A}$. We may define the anomaly $\mathcal{A}$ the non vanishing mean value $< \hat{A} >$. In many cases, this kind of anomaly may be evaluated non perturbatively in the following way[11].

Let $W[\alpha]$ the connected generating functional in the presence of a classical suitable source defined by

$$W[\alpha] = -\ln \int \mathcal{D}(\phi)\, e^{-(S[\phi] + \int \alpha A(\phi)\, dvol)} \tag{3}$$

Assuming that the observable A has a bilinear structure with respect to the fields ϕ, a new operator $H(\alpha)$ may be introduced, its general structure being $H(\alpha) = H + O(\alpha)$. It is defined by

$$W[\alpha] = -\ln \int D(\phi) e^{-\int \phi H(\alpha) \phi \, dvol} \tag{4}$$

This expression is formally related to the determinant of the operator $H(\alpha)$[12]. This can be easily seen by choosing a complete set of eigenvectors ϕ_n of $H(\alpha)$, that is

$$H(\alpha)\phi_n = \lambda_n \phi_n \tag{5}$$

In this way one has $\phi = \sum c_n \phi_n$. By definition, the diffeomorphism invariant measure is given by $D(\phi) = \prod (\mu/\sqrt{4\pi}) dc_n$, μ being a dimensional normalisation constant. From this it follows

$$W[\alpha] = -\frac{1}{2} \sum \ln \left(\frac{\lambda_n}{\mu^2} \right) = -\frac{1}{2} \ln \det \frac{H(\alpha)}{\mu^2} \tag{6}$$

which is valid for neutral scalar fields. Eq.(6) can be extended to any kind of fields (also anticommuting spinor fields [13]). In general one get

$$W[\alpha] = -p \ln \det \frac{H(\alpha)}{\mu^2} \tag{7}$$

where the numerical factor p depends on the nature of the fields involved. In particular, for spinor fields one has $p = -1$. From eq.(3), we also obtain an expression for the anomaly, namely

$$\mathcal{A} = \left. \frac{\delta W[\alpha]}{\delta \alpha(x)} \right|_{\alpha=0} \tag{8}$$

Another interesting characterisation of anomaly in terms of path integral is due to Fujikawa [14,15,16]. In that approach, the anomaly arises as non invariance of functional measure with respect to symmetry transformations. In fact in the path integral formulation, quantum effects are taken into account by the functional measure and in the presence of a symmetry invariance of the classical action, the only place where the quantum symmetry breaking can take place is the functional measure. To be more precise, let us start from the identity

$$Z[\mathcal{V}] = \int D(\phi) e^{-S[\phi, \mathcal{V}]} = \int D(\phi') e^{-S[\phi', \mathcal{V}]} \tag{9}$$

where $\phi' = \phi + \varepsilon(x)\phi + o(\varepsilon)$, ε being an infinitesimal gauge parameter. If we assume invariance of classical action, we have at first order in ε

$$S[\phi + \delta\phi, \mathcal{V}] = S[\phi, \mathcal{V}] - \int dvol \frac{\delta S}{\delta \mathcal{V}} \delta \mathcal{V} \tag{10}$$

where $\delta \mathcal{V} = D\varepsilon$, with D a suitable operator. Since $D(\phi') = D(\phi) \; J(\varepsilon, \mathcal{V})$, J being the functional Jacobian, we arrive at

$$Z[\mathcal{V}] = (1 + \delta J) \int D(\phi) e^{-S[\phi, \mathcal{V}]} \left(1 - \int dvol \frac{\delta S}{\delta \mathcal{V}} D\varepsilon \right) \tag{11}$$

which is valid again to the first order in ε . This gives

$$\delta J(\varepsilon, \mathcal{V}) = \int dvol < D^T \frac{\delta S}{\delta \mathcal{V}} > \varepsilon(x) =^{def} \int dvol\, \mathcal{A}(x)\varepsilon(x) \tag{12}$$

in which the action of D has been transferred on the "current". As a result, the anomaly has been associated to the non triviality of the functional Jacobian, which has the formal expression

$$\ln J(\varepsilon, V) = \ln \frac{\det H}{\det H(\varepsilon)} \tag{13}$$

where $\phi' H \phi' = \phi H(\varepsilon)\phi$ has been set.

In order to compute the anomaly, we have to give a precise meaning to the formal equations (7)and (13), which contain determinants of differential operators. To our p urposes, it is convenient to use the so called zeta function regularization [17,12], which can be briefly explained in the following way.

Let us introduce the zeta function of H as trace of the complex power H^{-s}, defined by the spectral theorem, that is

$$Tr\, H^{-s} = \sum_n E_n^{-s} \tag{14}$$

E_n being the non-zero eingenvalues of H. This series converges for $\Re s$ large enough, but can be analitically extended to a meromorphic function in the whole complex plane. Now

$$\lim_{s \to 0+} \frac{d}{ds} Tr H^{-s} = -\sum_n \ln E_n = -\ln \prod_n E_n = -\ln \det H \tag{15}$$

so we can define

$$\ln \det H = -\lim_{s \to 0+} \frac{d}{ds} Tr H^{-s} \tag{16}$$

In this way we get

$$W[\alpha] = -p \lim_{s \to 0+} \frac{d}{ds}(Tr\, H^{-s}(\alpha)/\mu^2) \tag{17}$$

$$\ln J(\varepsilon, \mathcal{V}) = \lim_{s \to 0+} \left(\frac{d}{ds} Tr \frac{H^{-s}}{\mu^2} - \frac{d}{ds} Tr \frac{H^{-s}(\varepsilon)}{\mu^2} \right) \tag{18}$$

Now, performing the functional variation of eq.(17), we arrive at

$$\delta W[\alpha] = \lim_{s \to 0+} \int dvol\, \alpha(x) tr[H^{-s-1}(\alpha)\delta H(\alpha)](x, x) \tag{19}$$

In most cases, the transformation law of H is linear homogeneous so that

$$\delta H(\varepsilon) = \delta\varepsilon X H + H Y \delta\varepsilon \tag{20}$$

in which X and Y are suitable matrices. This being so, an easy calculation ends with the result

$$\mathcal{A}(x) = \lim_{s \to 0+} tr[(X + Y) H^{-s}(x, x)] \tag{21}$$

Starting from eq.(18) the same result is easily achieved.

3 Heat kernel expansion and spectral coefficients

Eq.(21) tell us how heat kernel methods enter into the game. In fact, from the spectral theorem end eq.(14) we have (P_0 is the projector onto sero modes subspace)

$$H^{-s}(x,x') = \frac{1}{\Gamma(s)}\int_0^\infty dt\, t^{s-1}[K_t(x,x') - P_0(x,x')] \tag{22}$$

where we have introduced the heat kernel $K_t(x,x') =< x \mid exp(-tH) \mid x' >$, which satifies the heat type equation

$$\begin{aligned} &(\partial_t + H)K_t = 0 \\ &\lim_{t\to 0^+} K_t(x,x') = \delta(x,x') \end{aligned} \tag{23}$$

For our purposes, the amazing thing is that one needs only to know an asymptotic expansion of K_t for t around 0. Now general theorems assure that such asymptotic expansion exists and its structure is [18] (for a d-dimensional manifold without boundary)

$$K_t(x,x) \simeq \frac{1}{(4\pi t)^{d/2}}\sum_{n=0}^\infty a_n(x,H)t^n \tag{24}$$

where the "spectral coefficients" $a_n(x,H)$ have a universal geometric interpretation and are also known as Seeley-DeWitt coefficients.

The use of expansion (24) togheter with eqs.(21) and (22) leads to the formula, valid for even dimensions

$$\mathcal{A}(x) = tr(X+Y)[\frac{1}{(4\pi)^{d/2}}a_{d/2} - P_0] \tag{25}$$

which relates the anomaly to a particular spectral coefficient and, as we shall see, allows a non perturbative evaluation of a very large class of anomalies. From the above considerations, one arrives at the conclusion that a possible anomaly is computable as soon as one knows the spectral coefficient $a_{d/2}$ and the matrices X and Y, which are related to the symmetry transformation. In the case of odd dimension manifolf without boundary only the term proportional to the trace of P_0 survives in eq.(25) and this does not contribute to the anomaly when one goes back to Minkowski space.

Now we briefly summarize a useful method[11] for computing the coefficients of the heat kernel expansion for a generic second order-elliptic, non negative differential operator H defined on a d-dimensional Riemann-Cartan manifold without boundary. The technique that we propose is quite similar to the one introduced by Schwinger [19] and DeWitt [20].

The asymptotic solution of the heat kernel may be written in the form[21,20]

$$K_t(x,x') \simeq \frac{1}{(4\pi t)^{d/2}}\exp\left(-\frac{g_{rs}(\xi)\xi^r\xi^s}{4t}\right)\sum_{n=0}^\infty b_n(\xi)t^n \tag{26}$$

where $b_n(\xi(x,x'))$ are bi-spinor or bi-tensor densities of weight $-(1/2)$ and ξ^k are Riemannian coordinates in the neighbourhood of the point x. Since we are mainly interested in the computation of $a_n(x)$, which are local invariant quantities, the use of normal Riemannian coordinates is justified and it will notably simplify the computation. The coefficients $a_n(x,H)$ are related to $b_n(\xi)$ by

$$a_n(x,H) = \lim_{\xi \to 0} b_n(\xi) \tag{27}$$

For convenience, now we write H in the covariant form

$$H = -[\Box + 2W^k \nabla_k + M(x)] \tag{28}$$

where M is a non differential operator, W^k a generic matrix, ∇_k the covariant derivative with connection ω_k and $\Box = g^{ij}\nabla_i\nabla_j$. We have

$$\nabla_k\psi = (\partial_k + \omega_k)\psi = (\partial_k + A_k + \frac{1}{2}\Gamma_k^{\mu\nu}\Sigma_{\mu\nu})\psi \tag{29}$$

We use Greek and Latin characters for anholonomic and holonomic indices respectively. A_k is a gauge (in general non abelian) potential, and $\Sigma_{\mu\nu}$ are the generators of the representation of the Lorentz group ($SO(d)$) to which the field ψ belongs. We assume $\Gamma_k^{\mu\nu}$ to be the components of a metric compatible Cartan connection. Using expansion (26) in eq.(22) we get the recurrence equations for the quantities $b_n(\xi)$. In this general case, they read

$$\begin{aligned} &[\xi^k(\partial_k + \tilde{\omega}_k) + n]b_n(\xi) = -Hb_{n-1}(\xi); \quad ; \quad n = 0, 1, \ldots \\ &\tilde{\omega}_k = \omega_k + W_k - \tfrac{1}{2}S_k \quad ; \quad b_{-1} = 0 \end{aligned} \tag{30}$$

Here, S_k is the torsion trace. These equations permit us to compute $b_n(\xi)$, order by order, and therefore the $a_n(x)$ expansion coefficients. The explicit computation of a_1 and a_2 can be found in ref.[11]. Here we simply write down the final results. That is

$$a_1(x,H) = \tilde{M}(x) + \frac{1}{6}\hat{R}(x) \tag{31}$$

$$\begin{aligned} a_2(x,H) &= \frac{1}{180}(\Box\hat{R} + \hat{R}^{ijrs}\hat{R}_{ijrs} - \hat{R}^{ij}\hat{R}_{ij}) \\ &+ \frac{1}{2}(a_1)^2 + \frac{1}{6}(\tilde{\Box} + S^k\tilde{\nabla}_k)a_1(x) + \frac{1}{12}\tilde{\Omega}^2 \end{aligned} \tag{32}$$

where

$$\begin{aligned} &\tilde{M} = M + \frac{1}{2}\nabla_k S^k + \frac{1}{4}S^2 - (\nabla_k W^k + W^2) \\ &\tilde{\nabla}_k f = \nabla_k f + [W_k, f] \\ &\tilde{\Omega}_{ij} = \partial_i\tilde{\omega}_j - \partial_j\tilde{\omega}_i + [\tilde{\omega}_i, \tilde{\omega}_j] \end{aligned}$$

while $\hat{R}$, $\hat{R}_{ij}$ and $\hat{R}_{ijrs}$ are the scalar curvature, Ricci and Riemann tensors related to the Levi-Civita connection.

As has been shown in ref.[22], the a_n spectral coefficients determine the asymptotic behaviour for large continuous eingenvalues of the spectral measure associated with the self-adjoint operator H and enter the sum rules related to it.

4 Some examples of anomalies

Here we are going to discuss some significative examples of anomalies. To begin with let us consider the so called chiral anomaly which is the consequence of the chiral symmetry breaking. It is known in fact that the classical action of a massless Dirac field in the presence of an external electromagnetic field is invariant under the global chiral symmetry transformation $\psi \rightarrow \exp(-\varepsilon\gamma^5\psi)$. In the quantisation process this symmetry breaks down and the conservation of the relative chiral current acquires an anomalous term $\mathcal{A}_{chir}$. For this model we get

$$S(\alpha) = \int [\overline{\psi}\not{D}\,\psi + \alpha(x)\,\partial_\mu(\overline{\psi}\gamma^5\gamma^\mu\psi)]\,d^4x \tag{33}$$

from which $X = Y = -\gamma^5$ easily follows. Here $\not{D} = i\gamma^\mu(\partial_\mu - ieA_\mu)$ is the Dirac operator with electromagnetic potential A_μ, γ^μ and γ^5 being Euclidean Dirac matrices. Using eq.(21) we obtain the well known result[1]

$$\mathcal{A}_{chir} = \frac{e^2}{16\pi^2}\epsilon^{\alpha\beta\mu\nu}F_{\alpha\beta}\,F_{\mu\nu} - tr\,2\gamma^5 P_0 \tag{34}$$

where $F_{\mu\nu} = \partial_\mu A_\nu - \partial_\nu A_\mu$ is the usual field strength.

Another example of anomaly, close to the previous one, comes out if one considers a massless Dirac field interacting with (non-abelian) vector and pseudo-vector potentials (non-abelian axial anomaly)[2].

The model is described by the action

$$S = \int \overline{\psi}\not{D}\,\psi\,d^4x \tag{35}$$

where now $\not{D} = i\gamma^\mu(\partial_\mu + V_\mu + \gamma^5 A_\mu)$, $V_\mu = -igV_\mu^a T_a$, $A_\mu = -igA_\mu^a T_a$, T_a being the generators of the gauge group in the given representation.

The classical action (35) is invariant with respect to the (infinitesimal) gauge transformations

$$\delta_V\psi = -\epsilon\psi;\quad \delta_V V_\mu = \partial_\mu\epsilon + [V_\mu,\epsilon];\quad \delta_V A_\mu = [A_\mu,\epsilon]; \tag{36}$$

$$\delta_A\psi = -\epsilon\gamma^5\psi;\quad \delta_A V_\mu = [A_\mu,\epsilon];\quad \delta_A A_\mu = \partial_\mu\epsilon + [V_\mu,\epsilon]; \tag{37}$$

Transformation (37) is broken at the quantum level and a non-abelian axial anomaly $\mathcal{A}_a$ appears. By a calculation similar to the one of the previous example we get $X = Y = -\gamma^5$. Then it follows

$$\begin{aligned}\mathcal{A}_a = & - \frac{1}{8\pi^2}tr\,\gamma^5 a_2(\not{D}^2)T_a \\ = & - \frac{1}{16\pi^2}tr\,\epsilon^{\alpha\beta\mu\nu}\Big[F_{\alpha\beta}F_{\mu\nu} + \frac{1}{3}G_{\alpha\beta}G_{\mu\nu} + \frac{32}{3}A_\alpha A_\beta A_\mu A_\nu \\ & - \frac{8}{3}(F_{\alpha\beta}A_\mu A_\nu + A_\alpha A_\beta F_{\mu\nu} + A_\alpha F_{\beta\mu}A_\nu)\Big]\,T_a\end{aligned} \tag{38}$$

[1]The chiral anomaly has been computed also on a Riemannian manifold [20,23] and on a Riemann-Cartan manifold[24,25,26] too.

[2]This model has been extensively studied using perturbative methods by Bardeen[27], who gives for the first time the form of the consistent non-abelian axial anomaly. The same results has been derived using the path integral approach by Balachandran et al[28]. Recently the extension to curved spaces has been considered too[29].

where $F_{\mu\nu} = \partial_\mu V_\nu - \partial_\nu V_\mu + [V_\mu, V_\nu] + [A_\mu, A_\nu]$ and $G_{\mu\nu} = \partial_\mu A_\nu - \partial_\nu A_\mu + [V_\mu, A_\nu] - [V_\nu, A_\mu]$.

It has to be noted that $\not{D}$ is not a self-adjoint operator and this cause some difficulties in the application of heat kernel techniques. These difficulties can be avoided by making an analytic continuation $A_\mu \to iA_\mu$ on the $\not{D}$ operator[28,30]. Such a continuation has to be understood in eq.(38). As one can directly check, eq.(38) satisfies the Wess-Zumino integrability conditions[31].

Another important anomaly appears when one quantises a massless scalar field on a curved manifold of arbitrary dimension d. The classical action reads

$$S = \int \frac{1}{2}\phi \left[-\Box + \xi\hat{R}\right] \phi \, dvol = \int \phi H \phi \, dvol \tag{39}$$

which is conformally invariant if one chooses $\xi = \frac{(d-2)}{4(d-1)}$ (see for example ref.[32]). As a conseguence, the trace of energy momentum tensor is vanishing in virtue of classical equation of motion. With this choice, the trace of the expectation value of energy-momentum tensor becomes

$$T = < T^j_j > = < \phi \left[-\Box + \xi\hat{R}\right] \phi > = < \phi H \phi > \tag{40}$$

In order to obtain the above result, the diffeomorphism invariant measure plays a crucial role[10]. Recalling eq.(28), we get $H(\alpha) = (1+\alpha)H$, whence $X = 1, Y = 0$. Using eq.(21) we obtain (for even d)

$$\mathcal{A}_{conf} = \frac{1}{(4\pi)^{d/2}} \, a_{d/2}(H) \tag{41}$$

Then we see that after quantization, the trace of the energy-momentum tensor is not vanishing at all, but it is equal to the so called $\mathcal{A}_{conf}$ trace (or conformal) anomaly. This means that the conformal invariance is broken by the quantization process. In particular, on a 2-dimensional manifold we have

$$\mathcal{A}_{conf} = \frac{\hat{R}}{24\pi} \tag{42}$$

As a last example we consider the conformal anomaly associated with the Polyakov bosonic string[33]. The classical energy-momentum tensor of the string has a vanishing trace because of conformal invariance. Also in this case the conformal invariance is broken and a conformal anomaly appears.

The action of the string can be written in the form[34]

$$S = \int \frac{\delta^{ab}}{\rho} \partial_a x^\mu \partial_b x_\mu + (\xi P \eta + P^\dagger \xi \, \eta) \, d^2x \tag{43}$$

where the conformal gauge $g_{ab} = \rho\delta_{ab}$ has been chosen and $P = \sqrt{\rho}\gamma^a\partial_a\rho$, γ^a being the 2-dimensional Dirac matrices. Here $\mu = 1, ..., D$, D being the dimension of space-time. Using eq.(42) we see that the bosonic contribution to the conformal anomaly is equal to $\hat{R}\, D/24\pi$.

Setting

$$\Psi = \begin{pmatrix} \eta \\ \xi^t \end{pmatrix} \tag{44}$$

it is possible to write the ghost partition function in the form

$$\mathcal{F} = \int \mathcal{D}(\Psi) \exp \int_M \Psi^\dagger \mathcal{P} \Psi d^2x \tag{45}$$

where

$$\mathcal{P} = \begin{pmatrix} 0 & iP^\dagger \\ -iP & 0 \end{pmatrix} \tag{46}$$

The conformal anomaly we are interested in is related to the Fujikawa Jacobian associated to the following infinitesimal Weyl transformation

$$\Psi \to \Psi' = \Psi + \alpha(x)\Lambda\Psi \tag{47}$$

in which

$$\Lambda = -\begin{pmatrix} 2 & 0 & 0 & 0 \\ 0 & 2 & 0 & 0 \\ 0 & 0 & -1 & 0 \\ 0 & 0 & 0 & -1 \end{pmatrix} \tag{48}$$

As a result we have[3]

$$\ln J_G(\alpha) = -\lim_{s\to 0} Tr(\mathcal{P}^{-s}\Lambda\alpha) = -\int_M d^2x\alpha(x) tr[\mathcal{K}_0(x,\mathcal{P})\Lambda] \tag{49}$$

where the Seeley kernel[35] $\mathcal{K}_0(x,\mathcal{P})$ is the value at $s=0$ of the analitically continued kernel $\mathcal{K}_s(x,\mathcal{P}) = < x|\mathcal{P}^{-s}|x >$ which exists for $\Re(s)$ sufficiently large. Since $\mathcal{P}$ is a first order hermitian operator in an even dimensional manifold, $\mathcal{K}_0(x,\mathcal{P}) = \mathcal{K}_0(x,\mathcal{P}^2)$ [15], whence

$$\ln J_G(\alpha) = \int_M d^2x\alpha(x)[2\mathcal{K}_0(x,P^\dagger P) - \mathcal{K}_0(x,PP^\dagger)] \tag{50}$$

A direct computation, using the result (42) for the conformal anomaly of the bosonic sector and eq.(50), gives

$$\ln J(\alpha) = (D-26)\int_M \frac{d^2x}{24\pi}\hat{R}(x) \tag{51}$$

which is the well known result[7].

5 Conclusions

We have seen that several ways to characterize and compute anomalies are at our disposal. We have examined in some detail the path integral approach. This approch has some advantages and one can use the powerful methods related to the heat kernel expansion and zeta function regularization. It

[3] In this specific problem the use of the Seeley method in order to compute the kernel of $\mathcal{P}^{-s}$ is particularly indicated. We recall that the Seeley kernel for $s \to 0$ is proportional to $a_{d/2}$.

is also worth mentioning the BRST cohomological approach [36,37,8] which allows computations in higher dimensional space-times. However it seems that the physical meaning of this quantum mechanical symmetry breaking still remains to be completely understood. In this respect, it is interesting to observe that the non invariance of functional measure is equivalent to non invariance of fermiom determinant. As a conseguence, the anomaly may be interpreted as 1-cocycle and the unitary operator $U(g)$, which implement static gauge tranformations, gives only a projective representation [38] and the quantum algebra of the constraints acquires an extension [39]. In this context, it has been proposed that it might be possible to quantize gauge theory within this framework. Nevertheless no conclusive results have been yet emerged.

References

[1] E.S. Abers and B.W. Lee - Phys. Rep., **9**, 1 (1973).

[2] S. Adler - Phys. Rev., **177**, 1426 (1969).

[3] J. Bell and R. Jackiw - Nuovo Cimento A, **60**, 47 (1969).

[4] D.J. Gross and R. Jackiw - Phis. Rev. D, **6**, 477 (1972).

[5] C. Bouchiat, J. Iliopoulos and Ph. Meyer - Phys. Lett. B, **38**, 519 (1972).

[6] G. t'Hooft - Phys. Rev. Lett., **37**, 8 (1976); Phys. Rev. D, **14**, 3432 (1976).

[7] M.B. Green, J.H. Schwarz and E. Witten - Superstring theory, vol. 1. Cambridge University Press (1987).

[8] B. Zumino, Y.S. Wu and A. Zee - Nucl. Phys. B, **239**, 477 (1984).

[9] R. Jackiw - Chern-Simons terms and cocycles in physics and mathematics. Quantum field theory and quantum statistics - vol. 2 - I.A. Batalin, C.J. Isham and G.A. Vilkovisky eds.; Adam Hilgher, Bristol (1987).

[10] K. Fujikawa - Path integral quantization of gravitational interactions - local symmetry properties. in Quantum Gravity and Cosmology. World Scientific, Singapore; H. Sato and T. Inami editors. (1986).

[11] G. Cognola and S. Zerbini - Phys. Lett. B, **214**, 70 (1988).

[12] S.W. Hawking - Comm. Math. Phys., **55**, 133 (1977).

[13] P. Ramond - Field theory. The Benjamin/Cummings publishing company, Lomdon (1981).

[14] K. Fujikawa - Phys. Lett., **42**, 1195 (1979); Phys. Rev. Lett., **44**, 1733 (1980); Phys. Rev. D, **21**, 2848 ; **22** 1499 (1980).

[15] R.E. Gamboa Saravi, M.A. Muschietti, F.A. Schaposnik and J.E. Solomin - Ann. Phys., **157**, 360 (1984).

[16] M. Reuter - Phys. Rev. D, **31**, 1374 (1985).

[17] J.S. Dowker and R. Critchley - Phys. Rev. D, **13**, 3224 (1976).

[18] M. Atiyah R. Bott and V.K. Patodi - Inven. Math., **19**, 279 (1973).

[19] J. Schwinger - Phys. Rev., **82**, 664 (1951).

[20] B. DeWitt - The dynamical theory of groups and fields. Gordon and Breach, New York (1965).

[21] S. Minashisundaram and A. Pleijel - Canad. J. Math., **1**, 242 (1949).

[22] G. Cognola, L. Vanzo and S. Zerbini - Phys. Lett. B, **223**, 416 (1989).

[23] P.B. Gilkey - J. Diff. Geometry, **10**, 601 (1975).

[24] W.H. Goldthorpe - Nucl. Phys. B, **170**, 307 (1980).

[25] H.T. Nieh and M.L. Yan - Ann. Phys., **138**, 237 (1982).

[26] Yu.N. Obukhov - Phys. Lett. B, **108**, 308 (1982).

[27] W.A. Bardeen - Phys. Rev., **184**, 1848 (1969).

[28] A.P. Balachandran, G. Marmo, V.P. Nair and C.G. Trahern - Phys. Rev. D, **25**, 2713 (1982).

[29] G. Cognola and P. Giacconi Phys. Rev. D, **39**, 2987 (1988).

[30] A. Adrianov and L. Bonora - Nucl. Phys. B, **233**, 232 (1984).

[31] J. Wess and B. Zumino - Phys. Lett. b, **37**, 95 (1971).

[32] N.D. Birrell and P.C.W. Davies - Quantum fields on curved space. Cambridge University Press (1982).

[33] A.M. Polyakov - Phys. Lett. B, **103**, 207 (1981).

[34] L. Vanzo and S. Zerbini - Phys. Lett. B, **214**, 51 (1988).

[35] R.T. Seeley - Am. Math. Soc. Prog. Pure Math., **10**, 172 (1967).

[36] L. Bonora and P.C. Ramusino - Phys. Lett. B, **107**, 87 (1981).

[37] L. Bonora, P. Pasti and M. Tonin - J. Math. Phys., **27**, 2259 (1986).

[38] L.D. Faddeev and S.L. Shatashvili - Teor. Math. Fizika, **60**, 206 (1984).

[39] S.G. Jo - Nucl. Phys. B, **259**, 616 (1985).

DYNAMIC SYMMETRIES AND SUPERSYMMETRIES IN NUCLEAR PHYSICS

F. Iachello

Center for Theoretical Physics, Sloane Laboratory,
Yale University, New Haven, Connecticut 06511

Abstract

The role played by dynamic symmetries and supersymmetries in nuclear physics is discussed. Experimental examples are shown. The implications to other fields of physics are briefly reviewed.

1. Introduction

The structure of nuclei, especially medium-mass and heavy nuclei, is rather complex. An important tool in its study is provided by the use of symmetries. Symmetry is a wide reaching concept that is used in physics in a variety of ways. The symmetries that are particularly useful in nuclear physics are <u>dynamic</u> symmetries. In this article, after a brief review of the concept of dynamic symmetry, I will discuss its use in nuclear structure.

In the last 10 years the concept of symmetry has been enlarged even further with the introduction of super- or graded symmetries. The use of supersymmetry to study properties of odd-even nuclei will also be briefly reviewed. The type of supersymmetry found in nuclei is the only one that has been experimentally verified so far. Since the supersymmetric partners are nucleons (fundamental fermions) and their associated Cooper pairs (composite bosons), one may speculate on whether or not this type is the only one that can be realized in physics. A brief account of these speculations will be presented.

2. Dynamic Symmetries

The word symmetry, from the greek *συμμετρον*, meaning well-ordered, well-proportioned, is used in physics in many ways. Originally it applied to the geometric structure of physical systems, such as molecules, or crystals. Later, the concept was used to describe other situations, as for example those occurring in kinematic (or fundamental) symmetries. Examples of these symmetries are Poincare' invariance in relativistic quantum field theories and rotational invariance in non-relativistic quantum mechanics. In addition, there are permutational symmetries, gauge symmetries, etc.. Here, I discuss only one type of symmetry, dynamic symmetry. This is a situation in which:

(i) The Hamiltonian describing the system can be constructed in terms of the elements of a Lie algebra, $\mathcal{G}$ (called spectrum generating algebra).

(ii) The Hamiltonian can be written in terms only of invariant (Casimir) operators of a chain of algebras, $\mathcal{G}\supset\mathcal{G}'\supset\mathcal{G}''\supset\dots$. The simplest and most common case is that in which H is linear in the invariants,

$$H = \alpha C(\mathcal{G}) + \alpha' C(\mathcal{G}') + \alpha'' C(\mathcal{G}'') + \dots \quad , \tag{2.1}$$

where $C(\mathcal{G})$ denotes generically a Casimir operator of $\mathcal{G}$. More complicated functions can be (and have been) constructed as mentioned below in the case of the hydrogen atom.

Obviously (ii) is a special case of (i). The main advantage of dynamic symmetries is that all properties of the system can be given in closed form in terms of quantum numbers. These closed forms are useful in analyzing experimental data especially in complex situations such as those encountered in nuclei. The eigenvalues of H are given by

$$E = \alpha\langle C(\mathcal{G})\rangle + \alpha'\langle C(\mathcal{G}')\rangle + \alpha''\langle C(\mathcal{G}'')\rangle + \dots \quad , \tag{2.2}$$

where $\langle C(\mathcal{G})\rangle$ denotes the expectation value of $C(\mathcal{G})$ in the appropriate representation of $\mathcal{G}$. The energy formula (2.2) implies a splitting but not mixing of the representations of $\mathcal{G}$.

3. Dynamic symmetries in atomic and particle physics

The oldest example of a dynamic symmetry is provided by the spectrum of the non-relativistic hydrogen atom[1]. Pauli, Fock and Bargmann showed that the Schrödinger equation for the hydrogen atom has a larger symmetry than that implied by rotational invariance. The Hamiltonian H can be written in terms of the Casimir operator of this algebra, O(4), as

$$H = -\frac{me^2}{2(G_2+1)} \quad . \tag{3.1}$$

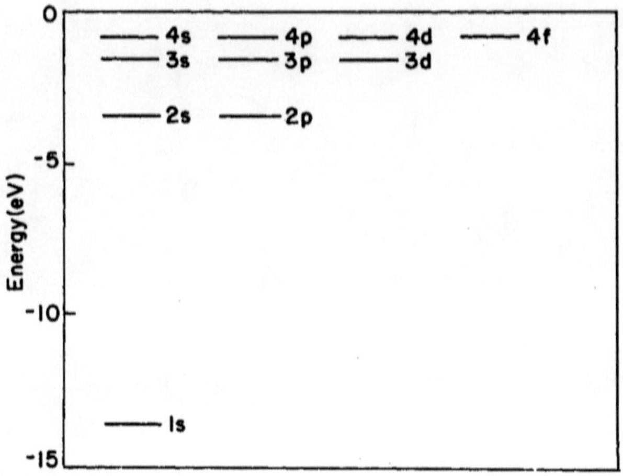

Fig.1. A portion of the level diagram of the non-relativistic hydrogen atom showing the regular pattern of energy levels and the occurrence of additional degeneracies.

Here G_2 is the first Casimir invariant of O(4). By taking the expectation value of (3.1) one obtains the Bohr formula

$$E(n) = -\frac{me^2}{2n^2} \quad , \qquad n = 1,2,\ldots \quad . \tag{3.2}$$

As one can see from Eq.(3.1), in this case it is actually 1/H and not H that is linear in the invariants. The presence of a dynamic symmetry manifests itself in: (i)the closed form (3.2) of the energy eigenvalues which produces a regular pattern of energy levels, as shown in Fig.1, and (ii)the occurrence of additional degeneracies.

Although known for at least 60 years, dynamic symmetries received considerable attention in physics only starting in the 60's when they were introduced in particle physics. In 1961, Gell-Mann and Ne'eman suggested[2] that the internal degrees of freedom of hadrons be described in terms of the Lie algebra of SU(3). This algebra is now called flavor SU(3) in order to distinguish it from other SU(3) algebras that play a role in hadronic physics (such as the color algebra). Furthermore, it was suggested that there exist a dynamic symmetry corresponding to the chain of algebras,

$$SU(3) \supset SU_I(2) \otimes U_Y(1) \supset O_I(2) \otimes U_Y(1) \quad , \tag{3.3}$$

where $SU_I(2)$ is the isospin algebra and $U_Y(1)$ the algebra of hypercharge. By writing the mass operator M describing the mass of the hadron as

$$M = M_0 + aC_1(U_Y(1)) + b\left[C_2(SU_I(2)) - (1/4)C_1^{\,2}(U_Y(1))\right] \quad , \tag{3.4}$$

and evaluating the expectation value of (3.4) in the representation

$$\left|\begin{array}{ccccccccc} SU(3) & \supset & SU_I(2) & \otimes & U_Y(1) & \supset & O_I(2) & \otimes & U_Y(1) \\ \downarrow & & \downarrow & & \downarrow & & \downarrow & & \\ (\lambda,\mu) & & I & & Y & & I_3 & & \end{array}\right\rangle \quad , \tag{3.5}$$

one obtains the Gell-Mann-Okubo[3] mass formula

$$M(I,I_3,Y) = M_0 + aY + b\left[I(I+1) - Y^2/4\right] \tag{3.6}$$

Eq.(3.6) provides an excellent description of hadronic spectra, as shown, for example, in Fig.2.

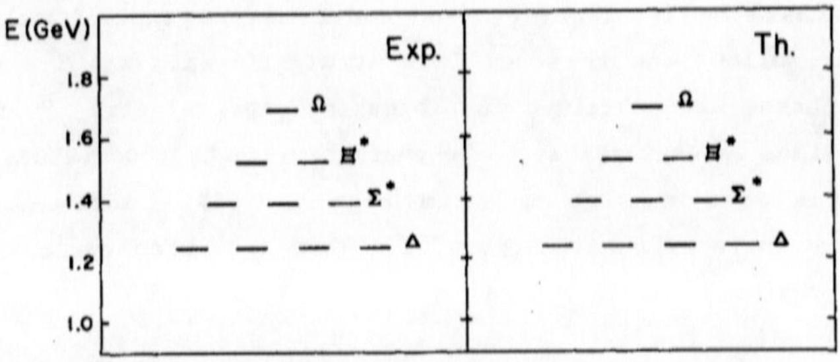

Fig. 2. Level diagram of the baryon decuplet. The energy levels are computed using the mass formula (3.6).

4. Dynamic symmetries in nuclear physics

Perhaps the best and most complete example of dynamic symmetries to date is provided by nuclear physics. Dynamic symmetries here are based on the interacting boson model, which I will now briefly review. In the interacting boson model[4], one assumes that low-lying collective states of nuclei with an even number of neutrons and protons can be constructed in terms of a system of N bosons with angular momenta $J^P=0^+$ or $J^P=2^+$. The bosons represent correlated pairs of neutrons and protons (Cooper pairs). It is convenient to introduce creation and annihilation operators for s(J=0) and d(J=2) bosons, denoted generically by $b^\dagger_\alpha$, b_α ($\alpha=1,\ldots,6$). The bilinear products

$$\mathcal{G} : \quad G_{\alpha\beta} = b^\dagger_\alpha b_\beta \qquad \alpha,\beta = 1,\ldots,6 \quad , \tag{4.1}$$

generate the Lie algebra of U(6). All operators are constructed from the operators $G_{\alpha\beta}$. For example, the Hamiltonian H is written as

$$H = E_0 + \sum_{\alpha\beta} \epsilon_{\alpha\beta} G_{\alpha\beta} + \frac{1}{2} \sum_{\alpha\beta\gamma\delta} u_{\alpha\beta\gamma\delta} G_{\alpha\beta} G_{\gamma\delta} \quad . \tag{4.2}$$

The algebra $\mathcal{G}$=U(6) is thus the spectrum generating algebra of this problem.

The fact that the interacting boson model has an explicit algebraic structure, U(6), allows one to study in a straightforward way its dynamic symmetries. These are obtained by breaking the algebra $\mathcal{G}$ in its subalgebras. Since we want states to be characterized by good values of the angular momentum, the angular momentum algebra, O(3), must always be contained in the chain of subalgebras of $\mathcal{G}$. There are then three and only three possible chains:

$$U(6) \begin{array}{l} \nearrow \\ \rightarrow \\ \searrow \end{array} \begin{array}{lr} U(5) \supset O(5) \supset O(3) \supset O(2) & (I) \;, \\ SU(3) \supset O(3) \supset O(2) & (II) \;, \\ O(6) \supset O(5) \supset O(3) \supset O(2) & (III) \;. \end{array} \tag{4.3}$$

In general, each nucleus is described by a set of parameters $\epsilon_{\alpha\beta}$, $u_{\alpha\beta\gamma\delta}$, and the problem of finding the eigenvalues of H must be solved numerically. However, for some nuclei, the values of the parameters are such that the Hamiltonian H can be written in terms only of the Casimir operators of one of the chains in (4.3). For example, for chain III,

$$H^{(III)} = E_0 + AC_2(O(6)) + BC_2(O(5)) + CC_2(O(3)) \quad , \tag{4.4}$$

where $C_2(\mathcal{G})$ denotes the quadratic invariant of the algebra $\mathcal{G}$. By evaluating the expectation values of H in the apprcpriate representation, one can then obtain the eigenvalues of H in terms of quantum numbers labelling the representations. In the case of chain III, the representations are labelled by

$$\left| \begin{array}{ccccccccc} U(6) & \supset & O(6) & \supset & O(5) & \supset & O(3) & \supset & O(2) \\ \downarrow & & \downarrow & & \downarrow & & \downarrow & & \downarrow \\ N & & \sigma & & \tau(\nu_\Delta) & & L & & M_L \end{array} \right\rangle \quad , \qquad (4.5)$$

and one obtains

$$E^{(III)} = E_0 + A\sigma(\sigma+4) + B\tau(\tau+3) + CL(L+1) \quad . \qquad (4.6)$$

Repeating the same procedure for the other chains one can obtain the full set of dynamic symmetry formulas,

$$E^{(I)}(N,n_d,v,n_\Delta,L,M_L) = E_0+\epsilon n_d+\alpha n_d(n_d+4)+\beta v(v+3)+\gamma L(L+1) \quad ,$$

$$E^{(II)}(N,\lambda,\mu,K,L,M_L) = E_0+\kappa(\lambda^2+\mu^2+\lambda\mu+3\lambda+3\mu)+\kappa' L(L+1) \quad , \qquad (4.7)$$

$$E^{(III)}(N,\sigma,\tau,\nu_\Delta,L,M_L) = E_0+A\sigma(\sigma+4)+B\tau(\tau+3)+CL(L+1) \quad .$$

These formulas can be used to analyze experimental data. In recent years several examples of nuclei that can be described by the three formulas (4.7) have been found. Three of these examples, one per each chain, are shown in Figs.3-5.

Since its introduction in 1974, the interacting boson model has been extended[4] to treat explicitly proton-neutron degrees of freedom, thus considering the algebra $U_\pi(6) \otimes U_\nu(6)$, called interacting boson model-2, and to consider proton-neutron pairs both with isospin one and zero (interacting boson-models-3 and 4)[5]. An analysis of the dynamic symmetries of these more complex models has also been done and some experimental examples found.

It should also be briefly remarked that, in addition to the particle

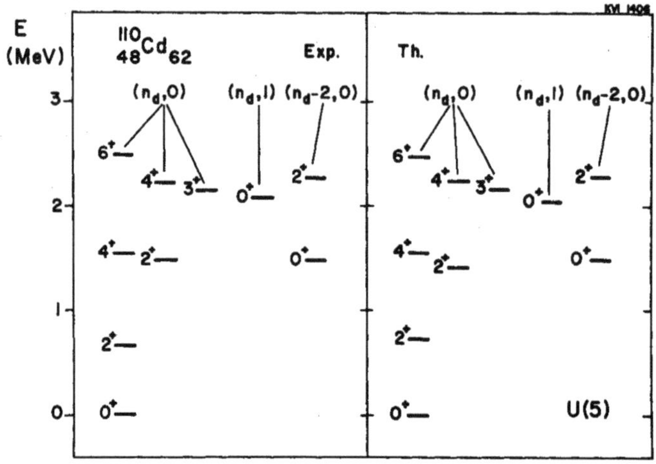

Fig. 3. An example of U(5) dynamic symmetry (chain I) in nuclei: ^{110}Cd. On the left the experimental spectrum. On the right the spectrum predicted by the energy formula (4.7),I.

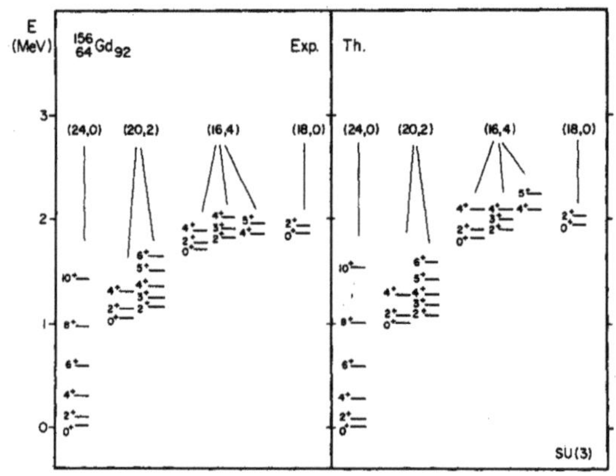

Fig.4. An example of SU(3) dynamic symmetry (chain II) in nuclei: ^{156}Gd. On the left the experimental spectrum. On the right the spectrum predicted by the energy formula (4.7), II.

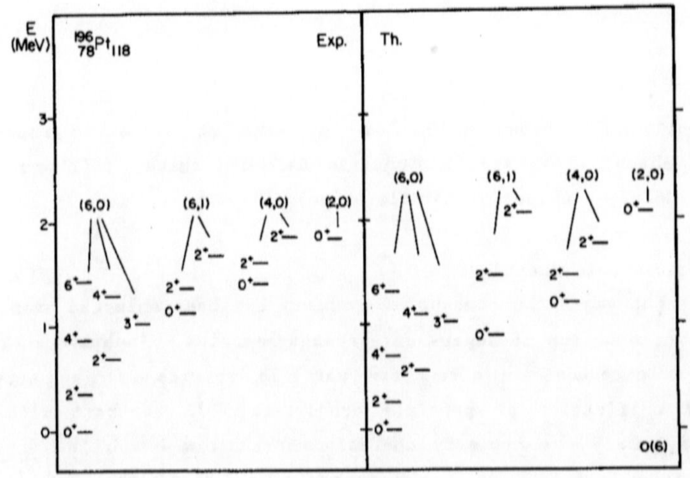

Fig. 5. An example of O(6) dynamic symmetry (chain III) in nuclei: ^{196}Pt. On the left the experimental spectrum. On the right the spectrum predicted by the energy formula (4.7),III.

interpretation of the bosons given above (Cooper pairs), it is possible to consider another (classical or geometric) interpretation. This is done through the introduction of a coherent or intrinsic state

$$|N,\alpha_\mu> = (s^\dagger + \sum_\mu \alpha_\mu d^\dagger_\mu)^N|0> \quad , \quad \mu = 0, \pm1, \pm2 \quad , \tag{4.8}$$

where the α_μ's are classical variables (c-numbers). The variables α_μ can then be associated with the shape of an object described by the surface

$$R = R_0(1 + \sum_\mu \alpha_\mu Y_{2\mu}(\theta,\phi)) \quad . \tag{4.9}$$

This leads to a description of collective states in terms of shape variables[6]. The three dynamic symmetries, Eq.(4.3), provide then a classification of shapes of nuclei in terms of symmetry groups: (I) spherical shape, (II) axially symmetric deformed shape, (III) non-axially symmetric deformed shape (γ-unstable shape).

5. Dynamic supersymmetries

In recent years the concept of symmetry has been enlarged even further with the introduction of super- or graded symmetries. Supersymmetries were originally introduced in elementary particle physics[7] in an attempt to provide a unification of particle properties. In contrast with normal symmetries, where the symmetry operations transform boson into bosons <u>or</u> fermions into fermions, in supersymmetry there are also operations that transform bosons into fermions and viceversa. As in the case of normal symmetries there are also here several types. Among these, particularly important are the kinematic (or fundamental) symmetries, as, for example, the Wess-Zumino[8] symmetry, and <u>dynamic</u> supersymmetries.

Dynamic sypersymmetries can be characterized by the following situation:

(i) The Hamiltonian describing the system is constructed in terms of the elements of a graded Lie algebra, $\mathcal{G}^*$ (called spectrum generating superalgebra).

(ii) The Hamiltonian can be written in terms only of invariant (Casimir) operators of a chain of algebras, $\mathcal{G}^* \supset \mathcal{G}'^* \supset \mathcal{G}''^* \supset \ldots$, i.e.

$$H = \alpha^* C(\mathcal{G}^*) + \alpha'^* C(\mathcal{G}'^*) + \alpha''^* C(\mathcal{G}''^*) + \ldots \quad , \qquad (5.1)$$

where $C(\mathcal{G}^*)$ denotes a Casimir operator of the algebra $\mathcal{G}^*$.

This definition of a dynamic supersymmetry is similar to that of a dynamic symmetry given in Sect.2, except for the appearance of graded algebras. The main difference between a normal and a graded Lie algebra is that in the graded algebras there appear two sets of operators, bosonic $G_\alpha(\alpha=1,\ldots,n)$, and fermionic $F_i(i=1,\ldots,m)$, satisfying the relations

$$\left[G_\alpha, G_\beta\right] = \sum_\gamma c^\gamma_{\alpha\beta}\, G_\gamma \quad ,$$

$$\left[G_\alpha, F_i\right] = \sum_j c^j_{\alpha i}\, F_j \quad , \qquad (5.2)$$

$$\left\{F_i, F_j\right\} = \sum_\alpha c^\alpha_{ij}\, G_\alpha \quad ,$$

together with the Jacobi identities. The presence in (5.2) of anticommutators makes the algebra a graded algebra. The importance of dynamic supersymmetries is that the eigenvalue problem for the Hamiltonian of a mixed system of bosons and fermions can be written explicitly in terms of quantum numbers, thus providing a simple classification scheme that can be checked easily by experiment. The eigenvalues of (5.1) are in fact obviously given by

$$E = \alpha^* \langle C(\mathcal{G}^*)\rangle + \alpha'^* \langle C(\mathcal{G}'^*)\rangle + \alpha''^* \langle C(\mathcal{G}''^*)\rangle + \ldots \quad , \qquad (5.3)$$

where $\langle C(\mathcal{G}^*)\rangle$ denotes the expectation value of $C(\mathcal{G}^*)$ in the appropriate representation of $G(\mathcal{G}^*)$.

6. Dynamic supersymmetries in nuclear physics

The use of dynamic supersymmetries in nuclear physics is based on the interacting boson-fermion model[9]. This model is a generalization of the interacting boson model and introduces, in addition to the bosons describing the collective Cooper pairs, fermions describing unpaired particles. The simplest example of nuclei where one must consider situations of this type is provided by nuclei with an odd number of protons and an even number of neutrons, or viceversa. In these nuclei, at least one particle must be unpaired. The fundamental ingredients in constructing operators that describe this situations are a set of boson creation and annihilation operators, $b^\dagger_\alpha$, $b_\alpha (\alpha=1,\ldots,n)$, as before, and a set of fermion creation and annihilation operators, $a^\dagger_i$, a_i $(i=1,\ldots,m)$. The elements of the Lie superalgebra can be written explicitly in terms of these operators as

$$G_{\alpha\beta} = b^{\dagger}_{\alpha}\, b_{\beta} \quad ,$$

$$G_{ij} = a^{\dagger}_{i}\, a_{j} \quad ,$$

$$F^{\dagger}_{\alpha i} = b^{\dagger}_{\alpha}\, a_{i} \quad ,$$

$$F_{\alpha i} = a^{\dagger}_{i}\, b_{\alpha} \quad , \tag{6.1}$$

where $G_{\alpha\beta}$ and G_{ij} are bosonic operators and $F^{\dagger}_{\alpha i}$ and $F_{i\alpha}$ are fermionic operators.

The Hamiltonian of the interacting boson-fermion model can be written as

$$H = H_B + H_F + V_{BF} \quad , \tag{6.2}$$

where H_B describes the bosons, H_F the fermions and V_{BF} their interaction. Each individual piece can be written in terms of the elements (6.1) of the superalgebra as

$$H_B = E_{OB} + \sum_{\alpha\beta} \epsilon_{\alpha\beta}\, G_{\alpha\beta} + \frac{1}{2} \sum_{\alpha\beta\gamma\delta} u_{\alpha\beta\gamma\delta}\, G_{\alpha\beta}\, G_{\gamma\delta} \quad ,$$

$$H_F = E_{OF} + \sum_{ij} \epsilon_{ij}\, G_{ij} + \frac{1}{2} \sum_{ijk\ell} u_{ijk\ell}\, G_{ij}\, G_{k\ell} \quad ,$$

$$V_{BF} = \sum_{\alpha\beta ij} w_{\alpha\beta ij}\, G_{\alpha\beta}\, G_{ij} + \sum_{\alpha\beta ij} w'_{\alpha\beta ij}\, F^{\dagger}_{\alpha i}\, F_{j\beta} \quad . \tag{6.3}$$

If one includes only s and d bosons and single particles in a shell with degeneracy Ω the spectrum generating superalgebra of (6.3) is the unitary superalgebra $U(6/\Omega)$.

Since there are three possible subalgebras of U(6) and many subalgebras

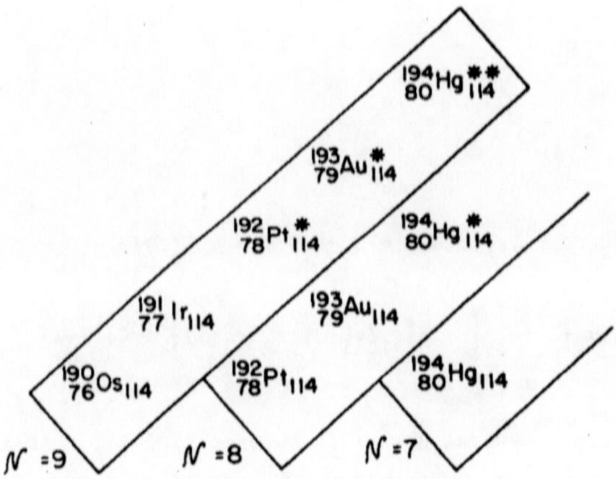

Fig. 6. U(6/4) supermultiplets in the Os-Ir region.

of U(Ω), depending on the actual degeneracy of the shell, there is a vast number of possibilities that can occur. Several of these possibilities have been investigated in detail. I discuss here only one case[10], that in which the single particle occupies a shell with $j=3/2$, for which $\Omega=2j+1=4$. The corresponding supersymmetry is U(6/4). By breaking the superalgebra into its maximal Lie subalgebra,

$$U(6/4) \supset U(6) \otimes U(4) \supset O(6) \otimes SU(4) \supset Spin(6) \supset$$
$$\supset Spin(5) \supset Spin(3) \supset Spin(2) \quad , \qquad (6.4)$$

one can then construct a dynamic supersymmetry. The eigenvalues of the Hamiltonian constructed in terms of invariants of (6.4) in the representation

$$\left|\begin{array}{ccccccccccccc} U(6/4) & \supset & U(6) & \otimes & U(4) & \supset & O(6) & \otimes & SU(4) & \supset & Spin(6) & \supset \\ \downarrow & & \downarrow & & \downarrow & & \downarrow & & & & \downarrow & \\ \mathcal{N} & & N & & M & & \Sigma & & & & \sigma_1,\sigma_2,\sigma_3 & \\ \\ \supset & Spin(5) & \supset & Spin(3) & \supset & Spin(2) & & & & & & \\ & \downarrow & & \downarrow & & \downarrow & & & & & & \\ & \tau_1,\tau_2,(\nu_\Delta) & & J & & M_J & & & & & & \end{array}\right\rangle \qquad (6.5)$$

are given by

$$E(\mathcal{N},N,M,\Sigma,\sigma_1,\sigma_2,\sigma_3,\tau_1,\tau_2,\nu_\Delta,J,M_J) = E_0+E_1N+E_2N^2+A\Sigma(\Sigma+4) +$$

$$+ A'\left[\sigma_1(\sigma_1+4)+\sigma_2(\sigma_2+2)+\sigma_3^{\,2}\right] + B\left[\tau_1(\tau_1+3)+\tau_2(\tau_2+1)\right] + CJ(J+1) \quad . \qquad (6.6)$$

This formula describes several nuclei (a supermultiplet), characterized by a given number of bosons plus fermions, $\mathcal{N}$. To each supermultiplet there belong five nuclei with $N=\mathcal{N}$, $M=0$; $N=\mathcal{N}-1$, $M=1$; $N=\mathcal{N}-2$, $M=2$; $N=\mathcal{N}-3$, $M=3$; $N=\mathcal{N}-4$, $M=4$. The quantum number M does not appear explicitly in Eq.(6.6) since it can be eliminated using the condition $N+M=\mathcal{N}$ and thus absorbed into E_0, E_1 and E_2. Typical supermultiplets are shown in Fig.6. The predicted spectra of a pair of nuclei (even-even and even-odd) are shown in Fig.7a. They can be compared with the experimental spectra shown in Fig.7b. Several other examples of supersymmetry have been found in the same region and in other regions of the periodic table.

7. Dynamic supersymmetries in particle physics

Dynamic supersymmetries of the type described above have also been considered in elementary particle physics. Indeed, the original suggestion of Miyazawa[7] was for a supersymmetry of this type. Recently, Catto and Gürsey[11] have suggested that the internal degrees of freedom of hadrons be described in terms of the superalgebra U(6/21). In this case, 6 is the dimension of the fermionic sector and 21 that of the bosonic sector. The

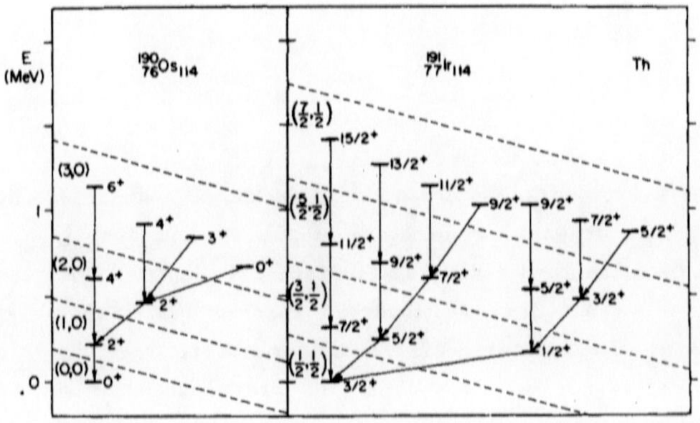

Fig. 7a. An example of U(6/4) supersymmetry in nuclei: (a) Excitation spectra predicted by Eq.(6.6).

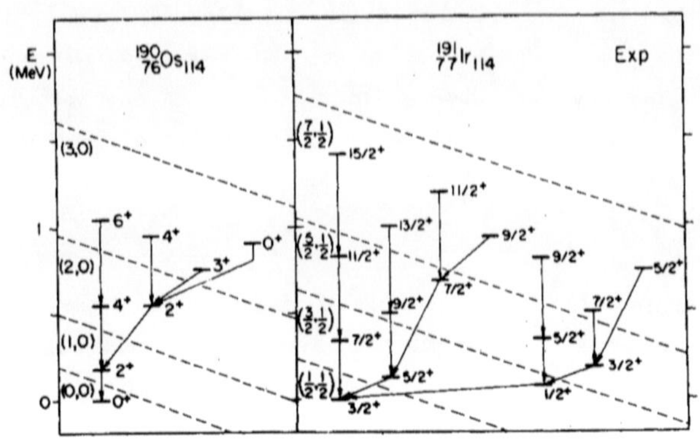

Fig. 7b. (b) Experimental spectra of the pair of nuclei ^{190}Os- ^{191}Ir.

supersymmetric partners are quarks and their associated Cooper pairs (diquarks). The fundamental supersymmetric multiplet is

$$\psi = \begin{pmatrix} q \\ \bar{D} \end{pmatrix} \quad , \tag{7.1}$$

where q denotes a quark and $\bar{D}$ an antidiquark. By constructing the mass operator in terms of Casimir operators of the appropriate algebras, one can obtain mass formulas that describe simultaneously baryons and mesons. These mass formulas account for the observed spectra quite well. One of the consequences of the supersymmetric mass formula is that the slope of the Regge trajectories for mesons and baryons are expected to be equal. This equality is experimentally observed.

Both supersymmetries described above are based on fundamental fermions and their associated Cooper pairs. Nambu[12] has recently suggested that these form a general class of supersymmetries which could eventually be applied in other situations such as those encountered in the electron gas or liquid ^{3}He. One can further speculate whether or not this type of supersymmetry is the only one that can be realized in physics. The experimental situation at the present time is in support of this speculation, since no example has been found so far of fundamental (Wess-Zumino) supersymmetries.

8. Conclusions

I have discussed here applications of the concept of dynamic symmetries to several situations in physics with particular emphasis to nuclear physics. The applications to nuclear physics are based on the interacting boson model. This model provides a unified framework within which properties of nuclei can be described. Although in this contribution I have concentrated my attention on the algebraic aspect of the model, it can be also related to the microscopic shell model[13]. From this point of view the interacting boson model combines together properties of the independent particle shell model and the collective model into a single framework. The model is able to describe and predict properties of complex nuclei with an accuracy of 10-20%.

9. Acknowledgements

This work was performed in part under U.S. D.O.E. Contract No. DE-AC-02-76-ER03074.

References:

1. W. Pauli, Z. Phys. 36, 336 (1926);
   V.A. Fock, Z. Phys. 98, 145 (1935);
   V. Bargmann, Z. Phys. 99, 576 (1936).
2. M. Gell-Mann, Phys. Rev. 125, 1067 (1962);
   Y. Ne'eman, Nucl. Phys. 26, 222 (1961).
3. S. Okubo, Progr. Theor. Phys. 27, 949 (1962).
4. F. Iachello and A. Arima, "The Interacting Boson Model", Cambridge University Press, Cambridge, 1987.
5. J.P. Elliott and A.P.White, Phys. Lett. 97B, 169 (1980);
   J.P. Elliott and J.A. Evans, Phys. Lett. 101B, 216 (1981).
6. A. Bohr and B.R. Mottelson, "Nuclear Structure", Vol. II, Benjamin, Reading, Mass. (1975).
7. H. Miyazawa, Progr. Theor. Phys. (Kyoto) 36, 1266 (1966);
   P. Ramond, Phys. Rev. D3, 2415 (1971).
8. J. Wess and B. Zumino, Nucl. Phys. B70, 39 (1974).
9. F. Iachello and O. Scholten, Phys. Rev. Lett. 43, 679 (1979).
10. A.B. Balantekin, I. Bars and F. Iachello,
    Phys. Rev. Lett. 47, 12 (1981); Nucl. Phys. A370, 284 (1981).
11. S. Catto and F. Gürsey, Nuovo Cimento 86A, 201 (1985).
12. Y. Nambu, Physica 15D, 147 (1985).
13. F. Iachello and I. Talmi, Rev. Mod. Phys. 59, 339 (1987).

Collective magnetic excitations in deformed nuclei

N. Lo Iudice

Dipartimento di Scienze Fisiche, Università di Napoli
and INFN, sezione di Napoli, Napoli (Italy)

1 Introduction

The variety of collective excitations occurring in all nuclei throughout the periodic table represents a major source of information for understanding nuclear structure.

The excitations are characterized by being strongly coupled to the ground state of the nucleus via an electric ($E\lambda$) or, to a less extent, a magnetic ($M\lambda$) multipole component of the electromagnetic (e.m.) radiation. Their transition strength $B(\lambda)$ is indeed several times larger than the single particle transition probability and can therefore be attributed only to a coherent motion of several nucleons.

In a semiclassical picture in which protons and neutrons are assumed to form classical fluids, these excitations are viewed as modes of oscillation of these fluids [1].

One distinguishes between isoscalar and isovector modes. In the first case no distinction is made between proton and neutron fluids, assumed to move in phase. The isovector modes are induced instead by a relative oscillation of proton versus neutron fluids.

The isoscalar excitations are low in energy with respect to the single particle excitation energies, while the isovector modes are considerably higher. This reflects the character of the nuclear forces in nuclei, which are attractive in the isoscalar channel and repulsive in the isovector one.

Let us suppose that the density oscillation is induced by an external field which in the case of isoscalar $E\lambda$ excitations is of the form

$$F_{\lambda\mu} = r^{\lambda} Y_{\lambda\mu}(\vec{r}) \ , \tag{1}$$

where $Y_{\lambda\mu}$ are spherical harmonics.
The density variation induced is to lowest order in the deformation parameters $\alpha_{\lambda\mu}$

$$\delta\rho = f_{\lambda}(r) \sum_{\mu} \alpha_{\lambda\mu} F^{+}_{\lambda\mu} \ , \tag{2}$$

where the $\alpha_{\lambda\mu}$ describe the density oscillation and are given apart from a constant factor by

$$\alpha_{\lambda\mu} = \int F_{\lambda\mu} \delta\rho d\vec{r} \ . \tag{3}$$

In the harmonic approximation the parameters $\alpha_{\lambda\mu}$ represent a set of $(2\lambda + 1)$ coordinates entering in the harmonic oscillator Hamiltonian

$$H = \frac{1}{2D_{\lambda}} \sum_{\mu} \pi^{+}_{\lambda\mu} \pi_{\lambda\mu} + \frac{1}{2} C_{\lambda} \sum_{\mu} \alpha^{+}_{\lambda\mu} \alpha_{\lambda\mu} \ , \tag{4}$$

where $\pi_{\lambda\mu}$ are the conjugate momenta and D_{λ} and C_{λ} are the mass and force constants respectively, to be determined by the use of some model, like the liquid drop model. The Hamiltonian can be written through a canonical transformation in terms of creation ($b^{+}_{\lambda\mu}$) and annihilation ($b_{\lambda\mu}$) boson operators and, apart from an additive constant, is

$$H = \omega_{\lambda} \sum_{\mu} b^{+}_{\lambda\mu} b_{\lambda\mu} \ , \tag{5}$$

where ω_{λ} is the frequency of the λ mode given by

$$\omega_{\lambda} = \sqrt{\frac{C_{\lambda}}{D_{\lambda}}} \ . \tag{6}$$

The first collective mode observed in nuclei was the celebrated electric dipole ($E1$) giant resonance. This is an isovector excitation which in the semiclassical limit is described as a relative translational oscillation between protons and neutrons [2]. The $\lambda = 1$ isoscalar collective excitation does not exist. This would correspond to a translational oscillation mode of the whole nucleus and would therefore not represent an intrinsic motion.

Another important electric collective mode, existing in both isoscalar and isovector channels, is the one described as a quadrupole ($\lambda = 2$) surface vibration around a spherical equilibrium shape.

The magnetic excitations, much less collective, are characterized by having a non-natural parity ($J^\pi = 1^+,\ 2^-,\ 3^+$, etc.). The excitation field in the isoscalar case is of the form

$$F_{\lambda\mu} = \vec{\sigma} \cdot \vec{\nabla}(r^\lambda Y_{\lambda\mu}) \ , \tag{7}$$

These modes were extensively studied not long ago in connection with the possible occurrence of spin-isospin phases in nuclei. It was in fact conjectured that, because of the strongly attractive pion-nucleon interaction, nuclear matter could undergo a transition from the normal to a spin-isospin ordered phase [3], called also pion condensation [4].

The proximity of such a phase in finite nuclei should have become manifest as a softening ($\omega \to 0$) of some magnetic collective mode and a dramatic enhancement of its corresponding $M\lambda$ transition probability [5].

The experimental analysis showed instead a strong quenching of the magnetic transitions. Apparently the repulsive short range spin-isospin dependent forces dominate over the long range pion-nucleon interaction.

It is to be noticed that the magnetic excitations studied, in particular the magnetic dipole modes, are assumed to be promoted by the spin dependent part of the magnetic operator (eq.(7)). In spherical nuclei indeed the magnetic dipole states can only be excited by the spin component of the M1 operator because of spherical symmetry. A collective $M1$ mode promoted by the orbital angular momentum component can occur however in deformed nuclei. This will be the main topic of this talk.

2 Orbital magnetic excitations: scissors mode

2.1 Motivation

Let us consider quadrupole surface vibrations [6]. These exist in spherical nuclei with few valence nucleons and represent oscillations around a spherical shape described by the set of five coordinates $\alpha_{2\mu}$. As one gets far from the closure of the shells, because of the attractive character of the force in this channel, the mode becomes soft and renders the nucleus unstable against deformation. When the number of valence nucleons reachs a critical value, a phase transition occurs with consequent change of the equilibrium shape, so that the nucleus from spherical becomes deformed. The symmetry is broken in the most economical way since the deformed shape preserves axial symmetry, as can be inferred from the rotational band structure of the spectra.

In deformed nuclei intrinsic collective quadrupole excitations still exist. They describe surface vibrations around the new equilibrium spheroidal shape. Because of the deformation the quadrupole modes are splitted into two collective excitations. The collective coordinates appropriate for describing these oscillations are the parameters $a_{2\nu}$ defined within the body-fixed frame. These can be obtained from the $\alpha_{2\mu}$ defined in the laboratory frame through the transformation

$$\alpha_{2\mu} = \sum_{\nu} a_{2\nu} D^{(2)}_{\mu\nu}(\Omega) \ , \tag{8}$$

where $\Omega \equiv \{\alpha\beta\gamma\}$ are the Euler angles describing the orientation of the system and $D^{(2)}_{\mu\nu}$ is the rotational matrix. Out of the five intrinsic components $a_{2\nu}$ only three describe an intrinsic collective motion. These are

$$\begin{aligned} a_{20} &= \beta \cos\gamma \\ a_{2\pm2} &= \beta \sin\gamma \end{aligned} \tag{9}$$

The first coordinate describes the so called β vibrations which preserve axial symmetry, the second describes γ vibrations which tend to break the above symmetry.

It is important to notice that the remaining components, $a_{2\pm1}$, are redundant. They would describe a rotation of the whole nucleus, already accounted for by Euler angles. This is the rotational analogue of the translational electric dipole mode discussed in the introduction. We can push, however, the analogy with the $E1$ case further and observe that, if distinction is made between protons and neutrons, the two fluids are allowed to rotate in opposition of phase around a common axis perpendicular to their symmetry axes and, because of the restoring neutron-proton force, can perform a relative rotational, scissors like, oscillation (Fig.1). This is the basic idea underlying the two-rotor model (TRM) [7].

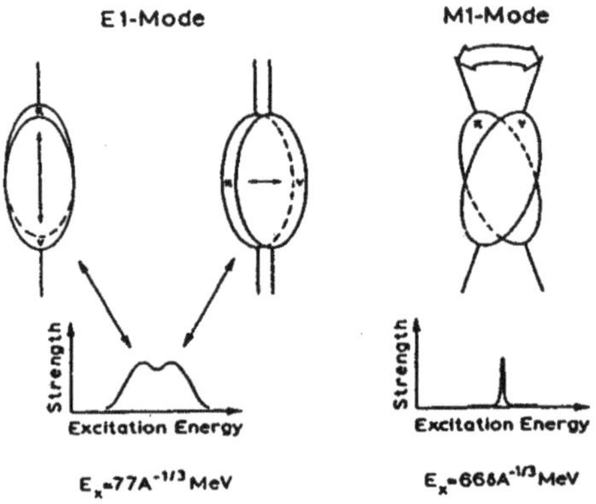

Fig.1. Classical two-fluid picture of collective vibrations of neutrons ν against protons π leading to isovector $E1$ and $M1$ modes, respectively.

2.2 The two-rotor model

The starting point of the TRM is a Hamiltonian describing two interacting rigid rotors, one for protons and one for neutrons, free to rotate around a common axis, which defines the intrinsic x_1 axis, orthogonal to their symmetry axes. This Hamiltonian is

$$H = \frac{1}{2\Im_p} J_p^2 + \frac{1}{2\Im_n} J_n^2 + V(\theta) \ , \tag{10}$$

where $\Im_p$ and $\Im_n$ are the proton and neutron momenta of inertia, $\vec{J}_p$ and $\vec{J}_n$ their angular momenta and $V(\theta)$ the potential describing the interaction between the two rotors, which depends on the angle 2θ between their symmetry axes.

As next step one has to separate the proton-neutron relative (intrinsic) motion from the rotation of the whole nucleus. To this purpose one defines the total and relative angular momenta

$$\vec{J} = \vec{J}_p + \vec{J}_n \quad , \quad \vec{S} = \vec{J}_p - \vec{J}_n \tag{11}$$

The Hamiltonian in terms of these new quantitities becomes

$$H = \frac{1}{2\Im}[J^2 + S^2] + \frac{\Im_n - \Im_p}{4\Im_n\Im_p}\,\vec{J}\cdot\vec{S} + V(\theta)\ , \tag{12}$$

where $\Im$ is essentially the moment of inertia of the whole nucleus and is given by

$$\Im = 4\frac{\Im_n\Im_p}{\Im_n + \Im_p}\ . \tag{13}$$

For small θ the Coriolis-like term can be neglected so that the Hamiltonian decouples into a rotational

$$H_{rot} = \frac{1}{2\Im}J^2 \tag{14}$$

and an intrinsic part. This, when expressed in terms of the realization of the intrinsic components of $\vec{J}$ and $\vec{S}$, in the harmonic approximation becomes

$$H_{int} = \frac{1}{2\Im}\left(-\frac{d^2}{d\theta^2} + \frac{J_3^2 - 1/4}{\theta^2}\right) + \frac{1}{2}C\theta^2\ . \tag{15}$$

The energy eigenvalues are

$$\omega_{nK} = \omega(2n + K + 1)\ , \tag{16}$$

where

$$\omega = \sqrt{\frac{C}{\Im}}\ . \tag{17}$$

The scissors mode corresponds to the first excited eigenvalue obtained by putting $n = 0$ and $K = 1$. These quantum numbers define a band of intrinsic excitation energy ω. Such a band is mainly excited through the orbital component of the magnetic dipole operator whose TR expression is

$$\mathcal{M}(M1,\mu) = \delta_{\mu,\pm 1}\sqrt{\frac{3}{32\pi}}S_1(g_p - g_n)\mu_N\ , \tag{18}$$

where g_p and g_n are respectively proton and neutron orbital gyromagnetic factors and μ_N is the nuclear magneton. The corresponding strength is

$$\begin{aligned} B(M1)\uparrow &= B(M1, 0^+ \to K^\pi = J^\pi = 1^+) \\ &= \frac{3}{16\pi}|\langle\varphi_1|S_1|\varphi_0\rangle|^2(g_p - g_n)^2\mu_N^2 \\ &\simeq \frac{3}{16\pi}\Im\omega(g_p - g_n)^2\mu_N^2\ , \end{aligned} \tag{19}$$

where we have denoted by φ_0 and φ_1 the TR ground and first excited intrinsic states respectively.

For a numerical estimate the restoring force constant was computed originally following a method similar to the one adopted by Goldhaber and Teller for the $E1$ giant resonance [2]. This method, however, leads to an overestimation of the energy and, to a greater extent, of the $M1$ strength with respect to the experimental findings. More realistic values can be obtained [8] by computing the restoring force constant through the energy mass formula

$$V(\theta) = k_{sym} \int \frac{(\rho_p(\theta) - \rho_n(-\theta))^2}{\rho^2} d\vec{r} = \frac{1}{2} C\theta^2 \,, \tag{20}$$

where $k_{sym} \simeq 28$ MeV is the symmetry energy constant, $\rho_p(\theta)$ and $\rho_n(-\theta)$ are the twisted proton and neutron densities and ρ the nuclear density.

The moment of inertia can be taken to have the rigid body value

$$\Im = \frac{2}{3} mA\langle r^2 \rangle \,. \tag{21}$$

Assuming for the density the Fermi distribution [9] we obtain for the frequency and the $M1$ strength

$$\omega \simeq 52\ \delta A^{-1/3}\ \text{MeV} \quad , \quad B(M1) \uparrow \simeq 0.04\ \delta A^{4/5} \mu_N^2 \,. \tag{22}$$

The estimate can be improved by taking into account the superfluid properties of the nucleus. This can be done by averaging the Inglis-Belajev formula of the moment of inertia over the single particle orbits, following the procedure of ref.[1]. The final result is

$$\omega \simeq 72\ \delta A^{-1/3}\ \text{MeV} \quad , \quad B(M1) \simeq 0.03\ \delta A^{4/3} \mu_N^2 \,. \tag{23}$$

2.3 Experimental evidence of the mode

An excitation with the properties of the $J^\pi = 1^+$, $K^\pi = 1^+$ state predicted by the TRM was discovered in a high resolution(e,e') experiment on ^{156}Gd [10]. It is now confirmed in three regions of the periodic table: the medium light nuclei, the rare earth region and the actinides [11]. The $M1$ excitation in ^{156}Gd was detected by performing (e,e') scattering at backward angles ($\theta = 165°$) where transverse magnetic transitions are dominant. Out of all $J^\pi = 1^+$

low lying levels only the strongly excited state at $E_x = 3.075$ MeV has the properties of the scissors mode. As shown in Fig.2 for some rare earth nuclei, the experimental form factor of this state is in close agreement with the TRM one, which, in the range of momentum transfert q shown, coincides with the form factor obtained within the interacting boson model (IBM), apart from a normalization constant. The IBM, however, reproduces better than the TRM the form factor at high q values, as shown for ^{164}Dy (Fig.3). We will discuss this point later on.

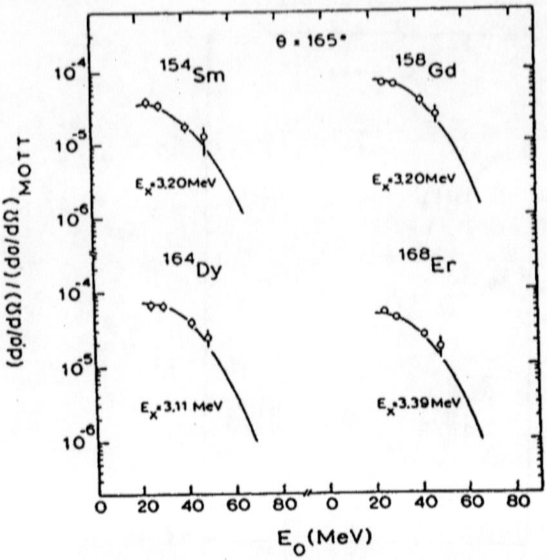

Fig.2. Magnetic dipole transition form factors of 1^+ states in various heavy deformed rare earth nuclei compared to IBM (TRM) predictions (from ref.[11]). In the range of q's shown the IBM and TRM form factors practically coincide.

This $M1$ state has been confirmed in (γ, γ') experiments [12]. In these reactions not only the elastic scattered photon was observed but also the decay of the $J^\pi = 1^+$ state to the $J^\pi = 2_1^+$ level of the ground state band. This allows to determine the K-value of the state under investigation using the Alaga rule.

In fact in the rotational limit we have

$$\begin{aligned} R &= \frac{B(M1, 1^+ \to 2_1^+)}{B(M1, 1^+ \to 0^+)} = \\ &= \frac{|\langle J^\pi = 2_1^\pi \ , \ K^\pi = 0^+ \| M(\lambda) \| J^\pi = 1^+ \ , \ K^\pi \rangle|^2}{|\langle J^\pi = 0_1^+ \ , \ K^\pi = 0^+ \| M(\lambda) \| J^\pi = 1^+ \ , \ K^\pi \rangle} = \\ &= \frac{|\langle 1 K \lambda - K | 2 \ 0 \rangle|^2}{|\langle 1 K \lambda - K | 0 \ 0 \rangle|^2} \end{aligned} \tag{24}$$

This rate is $R = 0.5$ if $K = 1$, $R = 2$ if $K = 0$. Experimentally R is about 0.5 consistently with the $K = 1$ assignement.

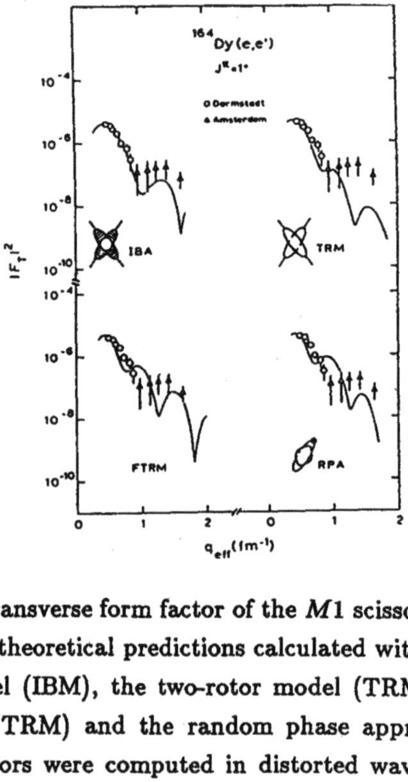

Fig.3. Transverse form factor of the $M1$ scissors mode in ^{164}Dy compared to theoretical predictions calculated within the interacting boson model (IBM), the two-rotor model (TRM), the fluid two-rotor model (FTRM) and the random phase approximation (RPA). All form factors were computed in distorted wave Born approximation (DWBA) and normalized to the data at the first maximum (from ref.[11]).

We have seen that the electron scattering form factors are in good agreement with the predictions based on the assumption that the excitation mode is of orbital nature.

A further support for this property of the mode is obtained by comparing the (e, e') with the (p, p') scattering. In (p, p') reactions in fact, the intermediate energy proton scattering at small angles excites magnetic dipole states only through the spin part of the nucleon-nucleon interaction.

The fact that the $M1$ states observed in electron scattering are not appreciably excited in the (p, p') reaction provides a strong evidence in favour of the orbital nature of the mode.

A more detailed experimental analysis shows that the mode is fragmented among several levels clustered around an energy, which in rare earth nuclei is about $E \simeq 3.1MeV$. Their summed $M1$ strength ranges from $B(M1) \uparrow \simeq 2\mu_N^2$ in ^{156}Gd to $B(M1) \uparrow \simeq 3.5\mu_N^2$ in ^{164}Dy.

The TRM accounts rather well for the energy ($E \simeq 3.4MeV$) and, as we saw, for the $M1$ form factor up to moderately high values of the momentum transfer. The model, however, because of its semiclassical nature, cannot account for the fragmentation of the mode. Moreover, its $M1$ strength, to be interpreted as a summed strength, is at least twice the corresponding experimental value. In rare earth nuclei in fact the TRM estimates gives $B_{TR}(M1) \uparrow \simeq 6.5\mu_N^2$.

For a more detailed and accurate desription more sophisticated approaches are needed.

3 Algebraic description of the mode: the interacting boson model

The basic assumption of the interacting boson model [13] is that the low-lying collective states of nuclei away from major closed shells are described in terms of a monopole boson with angular momentum and parity, $J^\pi = 0^+$, called s boson, and a quadrupole boson with $J^\pi = 2^+$, called d boson. These s and d bosons are interpreted as strongly correlated pairs of valence nucleons coupled respectively to $J^\pi = 0^+$ and $J^\pi = 2^+$.

In the first formulation of the model (IBM-I) no distinction is made between protons and neutrons. The Hamiltonian has the form

$$H_B = E_0 + \sum_{\alpha\beta} \varepsilon_{\alpha\beta} b_\alpha^+ b_\beta + \frac{1}{2} \sum_{\alpha\beta\gamma\delta} u_{\alpha\beta\gamma\delta} b_\alpha^+ b_\beta^+ b_\delta b_\nu \ , \qquad (25)$$

where E_0 is a c-number, $\varepsilon_{\alpha\beta}$ and $u_{\alpha\beta\gamma\delta}$ are free parameters, b_α^+ and b_α are respectively creation and destruction (s or d) boson operators. The distinguishing property of this boson Hamiltonian is that it contains in addition to the harmonic term (5) a two-body part and commutes with the total number of bosons. Moreover it has a definite group structure since the bilinear products are the 36 generators of the group $U(6)$. This allows to find the proper basis states for its diagonalization. In fact, one can generate from $U(6)$ three subgroup chains containing the rotation group as a subgroup. These are

$$U(6) \supset \begin{array}{ll} \nearrow \; U(5) \supset 0(5) \supset 0(3) \supset 0(2) & \mathrm{I} \\ \rightarrow \; SU(3) \supset 0(3) \supset 0(2) & \mathrm{II} \\ \searrow \; 0(6) \supset 0(3) \supset 0(2) & \mathrm{III} \end{array} \qquad (26)$$

Each group chain provides the quantum number for labeling a set of basis states.

The boson Hamiltonian, in its general form, can be diagonalized in one of these basis only numerically. For an appropriate choice of the parameters, however, it can be written as a sum of linear and quadratic Casimir invariants of one of the three chains. In these limiting cases the eigenvalue problem can be solved algebraically.

The states obtained in these three limits can be interpreted geometrically by making use of the concept of coherent states. It can be shown that in the chain I the states describe the vibrational motions of spherical nuclei, chain II corresponds to an axially symmetric rotor and chain III to γ-unstable rotor.

In the interacting boson model II (IBM-II) one makes distinction between neutrons and protons and assumes that the low lying levels of nuclei can be explained in terms of pairs of protons and pairs of neutrons with $J^\pi = 0^+$ and $J^\pi = 2^+$, treated as $s_\pi d_\pi$ and $s_\nu d_\nu$ bosons.

In order to take into account the particle-hole conjugation in particle space, the number of protons, N_π, and the number of neutrons, N_ν, is counted from the nearest closed shell.

The Hamiltonian for the coupled system of proton and neutron bosons can be written as

$$H = H_\pi + H_\nu + V_{\pi\nu} \ , \tag{27}$$

where $H_\pi (H_\nu)$ are the proton (neutron) boson Hamiltonian of the form (25) and $V_{\pi\nu}$ is the proton-neutron boson interaction.

Also the IBM-II has a group structure, that of the group product $U_\pi(6) \otimes U_\nu(6)$. One can generate from this group several group chains containing the rotation group as a subgroup, each one leading to a definite classical limit.

The chain leading to the scissors mode is the following

$$U_\pi(6) \otimes U_\nu(6) \supset U_{(\pi+\nu)}(6) \supset SU_{(\pi+\nu)}(3) \supset 0(3) \supset 0(2) \ . \tag{28}$$

Since the proton and neutron bosons are non identical, from the tensor product of the two $U(6)$ symmetric representations one generates symmetric as well as mixed symmetry representations. In order to label the states of different $U_{\pi+\nu}(6)$ symmetry it has become customary to introduce a new quantum number, called the F-spin, formally equivalent to isospin for particles. A proton boson has $F = 1/2$ and third component $F_3 = +1/2$, a neutron boson has $F = 1/2$ and $F_3 = -1/2$. For a system of $N = N_\pi + N_\nu$ bosons, the totally symmetric states have maximum F-spin, $F = F_{mx} = (N_\pi + N_\nu)/2$, the states with mixed symmetry have $F = F_{mx} - 1$ etc.

The most general IBM-II Hamiltonian has too many parameters to be of any practical use. A schematic Hamiltonian frequently used is of the form

$$H = \epsilon(n_{d\pi} + n_{d\nu}) + 2K_{\pi\nu}Q_\nu \cdot Q_\pi + K_{\nu\nu}Q_\nu \cdot Q_\nu + K_{\pi\pi}Q_\pi \cdot Q_\pi + \lambda M \ , \tag{29}$$

where $n_{d\pi}$ and $n_{d\nu}$ are proton and neutron $J^\pi = 2^+$ number operators, Q_π and Q_ν are the proton and neutron quadrupole operators and M is the so called Majorana operator related to the Casimir invariant of the group $U_{\pi+\nu}(6)$ and is responsible for the splitting between totally symmetric and mixed symmetry representations.

In the special case $\epsilon = 0$, and $K_{\pi\nu} = K_{\pi\pi} = K_{\nu\nu} = K$ the Hamiltonian (29) can be expressed as the sum of the Casimir invariants of the group chain (28) and is therefore diagonal in this scheme. The energy spectrum can be calculated algebraically. One obtains symmetric as well as mixed-symmetry bands, where $J^\pi = 1^+$ levels appear.

The first excited $J^\pi = 1^+$ state describes the scissors mode [14]. Due to the lack of experimental information on other mixed symmetry levels, the Majorana parameter is fixed by fitting the observed energy of this $J^\pi = 1^+$ state. The IBM cannot therefore predict the energy of the scissors mode.

The F-spin vector components of the $M1$ operator has the form

$$\mathcal{M}(M1,\mu) = \sqrt{\frac{3}{4\pi}}(g_\pi - g_\nu)\left(\frac{N_\nu}{N}L_{\pi,\mu} - \frac{N_\pi}{N}L_{\nu,\mu}\right) , \tag{30}$$

where N_π, N_ν are the neutron and proton pairs and $N = N_\nu + N_\pi$

The g_π and g_ν factors can be computed microscopically by a boson-fermion mapping procedure or, alternatively, can be estimated empirically from the comparison with the experimental g-factor of the 2^+ states. The values so determined are approximately $g_\pi \cong 1$ and $g_\nu \cong 0$.

The F-spin vector term couples the ground state to the $J^\pi = 1^+$ mixed symmetry state of the scissors mode with a strength given by [14], [15]

$$B(M1) = \frac{3}{4\pi}\frac{8N_\pi N_\nu}{2N-1}(g_\pi - g_\nu)^2\mu_N^2 . \tag{31}$$

It is to be noticed that only valence nucleons contribute to the strength. Using the values $g_\pi = 1$ and $g_\nu = 0$ for ^{156}Gd, $N_\pi = 7$ and $N_\nu = 5$, one obtains $B(M1) \cong 2.8\mu_N^2$, a value remarkably close to experiments.

As already mentioned the IBM form factor reproduces satisfactorily the experimental one up to the high momentum transfers (Fig.3), where the TRM appears to be inadequate. The IBM description of the mode is on the other hand closely related to the TRM. In the classical limit the IBM Hamiltonian assumes indeed the TRM form [16]. The success of the IBM over the semiclassical description was generally attributed to the fact that in IBM the neutron-proton rotational oscillation is performed by valence nucleons only (Fig.4). In order to check this point we modified the TRM by assuming that, just as illustrated in the figure, only the part of proton and neutron fluids external to an inert core takes part to the motion [8]. Such a change, however, spoils the agreement of the TRM with the experimental form factor (Fig.3). The good agreement of IBM is then to be ascribed to dynamical correlations and to spin degrees of freedom which the model accounts for in an effective way.

Although accounting satisfactorily for the $M1$ strength and form factor, the IBM, because of its phenomenological nature, has some limitations. Aside from not predicting the energy in fact, the model cannot account for the fragmentation of the mode. This can be hopefully explained only within a fully microscopic context.

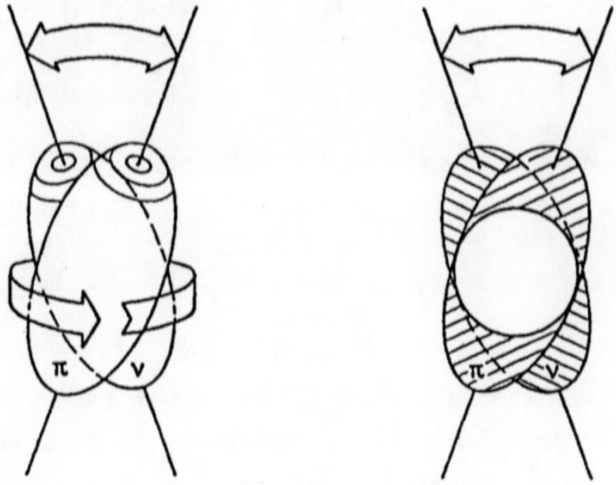

Fig.4. Classical TRM picture of the isovector $M1$ mode and description of the same mode within a modified TRM model, where a two-fluid picture of the rotors is assumed.

4 Microscopic Formulation

The link of the TRM with the microscopic approaches can be established in a quite general way [17], [18]. In the adiabatic approximation the total intrinsic wave function of the system can be written

$$\Psi_{\sigma K,n}(\theta, q) = \varphi_{\sigma K}(\theta)\phi_{n,\theta}(q) \ , \tag{32}$$

where the TR wave function $\varphi_{oK}(\theta)$ depends also on the quantum numbers n of the microscopic wave function $\phi_{n,\theta}(q)$, which has the twisted form

$$\phi_{n,\theta}(q) = e^{-i\theta S_1}\phi_n(q) \ . \tag{33}$$

The operator $\vec{S}$ is still defined as in eq.(11), but now $\vec{J}_p$ and $\vec{J}_n$ are shell model one-body operators. The classical TR potential is obtained as the zero-point energy of the twisted ground state with $n = 0$. Such a potential, to leading order in θ, is

$$V(\theta) = \Delta E = E_\theta - E_o \simeq \frac{1}{2}C\theta^2 \ , \tag{34}$$

where

$$C = <\phi_o|[S_1,[H,S_1]]|\phi_o> = 2\sum_n \omega_n| < \phi_n|S_1|\phi_o > |^2 \ . \tag{35}$$

The moment of inertia can be defined by extending the Inglis formula to the case of isovector rotations. To this purpose one can put in eq.(33) $\theta = i\omega t$ and define

$$\Im = \frac{1}{\omega} < \varphi_\theta|S_1|\varphi_\theta > \simeq 2\sum_n \frac{| < \phi_n|S_1|\phi_o > |^2}{\omega_n} . \tag{36}$$

The above equations define the TR quantities completely in microscopic terms.

We are now able to study the microscopic content of the TRM and to ascertain its limitations. Let us consider the $M1$ transition which represents the signature of the scissors mode.

If the spin contribution can be neglected, the isovector $M1$ operator has the form:

$$\mathcal{M}(M1,\mu) = \left(\frac{3}{16\pi}\right)^{1/2} S_\mu \mu_N \ , \tag{37}$$

where

$$S_\mu = \sum_i \tau_3(i) l_\mu(i) = \sum_i l^p_\mu(i) - \sum_i l^n_\mu(i) \ . \tag{38}$$

In this approximation the following energy weighted sum rule holds

$$\sum_n \omega_n B_n(M1) \uparrow = \frac{3}{8\pi}\sum_n \omega_n| < \varphi_n|S_{+1}|\varphi_o > |^2\mu_N^2 = \frac{3}{16\pi}\Im\omega^2\mu_N^2 \ . \tag{39}$$

The TRM as most schematic models correspond to the extreme limit in which the full strength is concentrated into a single state of frequency ω, so that

$$B(M1) \uparrow = \frac{3}{8\pi}| < \phi_1|S_{+1}|\phi_o > |^2\mu_N^2 = \frac{3}{16\pi}\Im\omega\mu_N^2 \ , \tag{40}$$

which is the TRM expression.

Such an ideal situation never occurs in practice. The operator S_{+1} excites in general several states. One could at most hope that the corresponding levels pack around the value $\omega = \sqrt{\frac{\Im}{C}}$. The following approximate relation would then hold

$$B_{TR}(M1) \uparrow \simeq \sum_n B_n(M1) \uparrow \simeq \frac{3}{8\pi} \sum_n | < \phi_n | S_{+1} | \phi_\circ > |^2 \mu_N^2 \simeq \frac{3}{16\pi} \Im \omega \mu_N^2 \; . \quad (41)$$

The TR expression of the $M1$ strength should then be considered as an approximation to the corresponding microscopic integrated strengths.

This situation does not correspond to reality either. As we shall see, only a group of levels cluster around ω, while other states are quite high in energy. Even in such a case the TRM can be made consistent with the microscopic descriptions, as long as the low and high energy spaces are decoupled. This can be achieved by simply truncating the space and defining effective operators in such a low energy model space. The effects due to the coupling with the high energy configurations can be taken into account by properly renormalizing these operators. With these limitations the microscopic integrated strength is still given to a good approximation by the TRM expression (41), where the sums are now limited to the low energy states.

An illustrative example of the present analysis is represented by schematic random-phase-approximation (RPA).

5 TRM and schematic RPA

The simplest and most natural RPA description of the mode can be given using a cylindrical harmonic oscillator basis and a quadrupole-quadrupole ($Q - Q$) two-body potential [17], [19], [21]. The single particle basis can be considered as obtained selfconsistently from this interaction. The residual $Q - Q$ interaction appropriate for describing the $M1$ state is:

$$V = K \sum_{i<j} Q^+(i) Q(j) \; , \quad (42)$$

where

$$Q = \tau_3 r^2 Y_{21} \; . \quad (43)$$

Such a Hamiltonian is suggested by the fact that the operators S and Q excite the same set of particle-hole $(p-h)$ states

$$\begin{aligned} S_{+1}|> &= \sum_{p-h} |p-h><p-h|S_{+1}|> \\ &= -(\frac{8\pi}{15})^{1/2} m(\omega_3^2-\omega_1^2) \sum_{p-h} |p-h> \frac{<p-h|Q|>}{\varepsilon_p-\varepsilon_h} , \end{aligned} \quad (44)$$

where ω_1 and ω_3 are respectively the transverse and longitudinal harmonic oscillator frequencies. These are related among themselves so that $\omega_1^2\omega_3 = \omega_0^3$, where ω_0 is the frequency at zero deformation.

Alternatively we can use the arguments of the unified theory and observe that the density variation induced by an isovector rotation around the x_1 axis is proportional to Q. In fact, to leading order in the deformation parameter, we get

$$\delta\rho = \delta\rho_p(\theta) - \delta\rho_n(-\theta) \simeq R\frac{d\rho}{dR} \sum_{\mu=\pm1} a_{2\mu}(\theta) Y_{2\mu}(\hat{r}) , \quad (45)$$

where

$$a_{2\mu}(\theta) \simeq a_{2\mu}^p(\theta) - a_{2\mu}^n(-\theta). \quad (46)$$

According to the unified theory [21], the semiclassical description of the mode in terms of the collective variable θ, or $a_{21}(\theta)$, is equivalent to a schematic RPA, with a $Q-Q$ interaction, as long as the $p-h$ excitations are quasi-degenerate. This is not the case, however [18]. The operators Q and S_1 excite not only the $0-\hbar\omega$ $p-h$ states of energy $\epsilon = |\omega_1-\omega_3| \simeq \delta\omega_\circ$, entering into the scissors mode, but also the $2-\hbar\omega$ $p-h$ states of energy $\varepsilon_Q = \omega_1+\omega_3 \simeq 2\omega_\circ$, which enter into the state describing the isovector quadrupole giant resonance.

The two approaches seem even less compatible since, at the unperturbed level, the strength of S_{+1} goes mostly to the low energy states, while that of Q is mainly collected by the $2-\hbar\omega$ states. We have in fact for the $E2$ unperturbed strength $B_\circ(E2)$:

$$B_{\circ,S}(E2)\uparrow = \frac{e^2}{2}\sum_S |Q_{p-h}|^2 \simeq \frac{\delta}{2} B_{\circ,Q}(E2) , \quad (47)$$

where:

$$B_{\circ,Q}(E2) = \frac{e^2}{2}\sum_Q |Q_{p-h}|^2 \simeq \frac{5}{16\pi}\frac{A<r^2>}{m\omega_\circ}e^2 . \quad (48)$$

The $M1$ strengths are given instead by

$$B_{o,Q}(M1)\uparrow = \frac{3}{8\pi}\sum_{Q}|(S_{+1})_{p-h}|^2\mu_N^2 \simeq \frac{\delta}{2}B_{o,S}(M1)\uparrow\ , \tag{49}$$

where

$$\begin{aligned} B_{o,S}(M1)\uparrow &\simeq \frac{3}{16\pi}\Im\omega_{o,S}\mu_N^2\ , \\ \omega_{o,S} &= \varepsilon \simeq \delta\omega_o\ . \end{aligned} \tag{50}$$

The two indices S and Q refer respectively to the $0-\hbar\omega$ and to the $2-\hbar\omega$ $p-h$ spaces. The scissors mode pertains to the first space, the isovector quadrupole mode to the other.

One could be tempted to neglect the high energy $p-h$ excitations, given the fact that they get little $M1$ strength. This, however, would be a poor approximation since the two spaces are coupled by the interaction. The importance of the $2-\hbar\omega$ space can be clearly seen from the fact that the low and high energy spaces give a comparable contribution to the TR restoring force constant. We have indeed from eqs.(35), (39), (49) and (50)

$$C = C_{0,S} + C_{0,Q} \simeq 2C_{0,S} \simeq \Im(\delta\omega_o)^2\ . \tag{51}$$

The apparent incompatibility between TRM and RPA is removed within the framework of the theory of the effective interaction. According to such a theory we can solve the RPA eigenvalue problem within the $0-\hbar\omega$ $p-h$ space and account for the coupling with the high energy space by defining properly renormalized effective operators.

In the present case the goal is reached by defining in the truncated space the same $Q-Q$ interaction (42) with a renormalized coupling constant K_{eff}. Once the problem has been confined within the low energy truncated space it is immediate to include the pairing corrections in the BCS approximation. In this approximation the pairing contribution is accounted for by turning the $p-h$ excitation energy ε into two quasi-particle energy E and by multiplying the $p-h$ matrix elements of the time-odd $M1$ operator by a factor

$$u_{n_1+1}v_{n_1} - u_{n_1}v_{n_1+1} \simeq v_{n_1}^2 - u_{n_1}^2 \simeq \frac{\epsilon}{E}, \tag{52}$$

where u^2 and v^2 are the hole and particle occupation probabilities. The single particle states are the cylindrical harmonic oscillator states with quantum numbers $[Nn_1\Lambda]$.

The $M1$ strength remains of the same form (50). We have in fact

$$B(M1)\uparrow \simeq \frac{3}{16\pi}\Im_{qp}\omega\mu_N^2 ,$$

where

$$\Im_{qp} \simeq \frac{1}{2}\Im_{rig} \quad \text{and} \quad \omega = E\left(1+\frac{K_{eff}}{K_0}\right) .$$

Here K_0 is the restoring force constant of the unperturbed excitation mode of energy E [20]. Numerically, one obtains [19]

$$\begin{aligned} \omega &\simeq 66\,\delta A^{-1/3} MeV , \\ B(M1)\uparrow &\simeq 0.03\,\delta A^{4/3}\mu_N^2 . \end{aligned} \tag{53}$$

These values are quite close to the ones obtained with the TRM. Consequently, also schematic RPA overestimates the $M1$ strength by a factor two. For a closer agreement to be reached one needs to exploit the shell structure more in details by performing realistic RPA calculations.

6 Microscopic calculations

Microscopic calculations are necessary for a detailed and complete account of the properties of the mode.

The mode, being a low energy excitation, is in fact fragmented. Such a fragmentation can be accounted for only by explicitely including the shell structure in the description.

In light and medium light nuclei the mode has been described successfully within the standard shell model scheme. A good account of the mode in Ti isotopes has been given, for instance, by Zamick [22] using a seniority basis, which can be considered to some extent as a microscopic realization of the IBM basis. Such a shell model basis in fact includes monopole and quadrupole pairing correlations as well as components of higher multipolarity.

In heavy nuclei full shell model calculations are not possible. The only feasible microscopic calculations are in practice of RPA type. A large variety of "realistic" RPA calculations have been indeed devoted to the study of the scissors mode [23]. Unfortunately these calculations not only did not converge to a unique conclusion but often gave contraddictory results.

Some of them, for instance [24], lead to results which question the collective as well as the "scissors" character of the mode, but on the other hand produce a form factor which is at variance with the experimental one (Fig.3). Some others give too much strength and too fragmentation [25]. The reason of these contraddictory RPA results can be understood if one observes [18] that the quantities characterizing the properties of the mode are quite sensitive to the Hamiltonian. This can be inferred from eqs.(35) and (39). A microscopic calculation should then take into consideration all the pieces of the Hamiltonian which most affect these quantities and treat them in the most appropriate basis. The success of the IBM description suggests that one should formulate the problem in the laboratory frame in order to avoid spurious admixtures of the mode with purely rotational excitations, should at least consider monopole and quadrupole pairing in addition to the quadrupole-quadrupole potential and should use a basis which privileges the pairing correlations over the single particle properties. The above requirements are not easily met in the RPA approach, which privileges the single particle properties and is naturally formulated in the intrinsic frame. Moreover the Hamiltonians generally used in actual calculations are incomplete and treated in an approximate fashion. Most of them in fact take into consideration the pairing interaction only, treated in the BCS approximation. A major progress towards a satisfactory description of the mode within a RPA scheme has been achieved in recent calculations, where the spurious states are disintangled from the scissor state either from the beginning by performing the RPA calculation in a self-consistent Hartree-Bogoliubov basis [26] or by adding counter-terms in the Hamiltonian [27]. Both calculations also include most of the relevant terms of the Hamiltonian, reaching satisfactory results. The agreement with experiments, including the form factor, appears quite satisfactory in the calculation of ref.[27], where the problem is formulated in an extended shell model space.

7 Concluding remarks

We have discussed only few of the many theoretical approaches used to describe the scissors mode. The large variety of calculations can be grouped into three main categories: i) Semiclassical descriptions, like the TRM; ii) algebraic descriptions, like the IBM; iii) microscopic descriptions, like RPA.

It is at first surprising that a single mode lends itself to be described in so many, apparently unrelated, approaches. The explaination is to be found in the fact that the kind of correlation giving rise to the mode, singled out in the TRM, is quite simple and can therefore be described in many different ways. The TRM correlation represents indeed the unifying feature of all different approaches. For the rest these can be quite different from each other and lead in general to quite different results. The only reasonable criterion for discriminating among these different descriptions is to compare their predictions with the observed properties of the mode.

In this respect we have seen that the TRM accounts reasonably well for the gross properties of the mode. We have also seen that, among the phenomenological descriptions, the one provided by the IBM is the most accurate, accounting satisfactorily for the $M1$ strength as well as for the $M1$ form factor up to high momentum transfer. Because of its phenomenologic nature, however, also the IBM description has its limitations: It does not predict the energy of the mode and cannot account for the fragmentation of the $M1$ strength.

For a detailed description of the properties of the mode a microscopic approach is needed. We have seen that the mode can be described naturally within different microscopic contexts.

In heavy nuclei, on the other hand, the only possible microscopic approach is of RPA type. We have pointed out, however, that this is not the most natural approach in many respects, as confirmed by the inadequacy of most of the RPA calculations carried out so far to describe the mode.

We have enumerated the limitations of these calculations and pointed out at the same time that, if certain prescriptions are fulfilled, as in recent calculations, the RPA approach can account satisfactorily for the detailed properties of the mode.

We can then say that we are approaching a stage in which the mode is understood not only semi-quantitatively, but also in its detailed properties.

We add as a final comment that the mode represents a beautiful manifestation of dynamical symmetries in nuclei, and that its study has allowed to gain a deeper understanding of the interplay between the single particle motion, which tends to break this symmetry, and the collective behaviour of nucleons which tend to enforce it.

References

1. For an exhaustive description of collective modes see: A. Bohr and B.R. Mottelson, "Nuclear structure", Benjamin, N.Y., 1975, vol.II.
2. M. Goldhaber and E. Teller, *Phys. Rev.* **74**, 1046 (1948).
3. F. Calogero, in "The Nuclear Many-Body Problem", ed. by F. Calogero and C. Ciofi Degli Atti, Editrice Compositori, Bologna, 1972, vol.2, p.535; F. Calogero and F. Palumbo, *Lett. Nuovo Cimento* **6**, 663 (1973).
4. A.B. Migdal, *Z. Exsp. Teor. Fiz.* **61**, 2209 (1971) [*Sov. Phys. JETP* **34**, 1184 (1972)].
5. For a review see for instance: E. Oset, H. Toki and W. Weise, *Phys. Rep.* **83**, 281 (1982).
6. See ref.1, chap.6.
7. N. Lo Iudice and F. Palumbo, *Phys. Rev. Lett.* **41**, 1532 (1978); G. De Franceschi, F. Palumbo and N. Lo Iudice, *Phys Rev.* **C25**, 1496 (1984).
8. N. Lo Iudice, F. Palumbo, A. Richter and H.J. Wortche, to be published.
9. See for instance: A. Bohr and B.R. Mottelson, "Nuclear structure", Benjamin, N.Y., 1969, vol.I.
10. D. Bohle, A. Richter, W. Steffen, A.E.L. Dieperink, N. Lo Iudice, F. Palumbo and O. Scholten, *Phys. Lett.* **B137**, 27 (1984).

11. For an updated review of the experimental situation see: A. Richter, in "Contemporary topics in nuclear structure physics", R.F. Casten et al. (eds.), World Scientific, Singapore, 1988, pp.127-164.

12. U.E.P. Berg, C. Blasing, J. Drexler, R.D. Heil, U. Kneissel, W. Naatz, R. Ratzek, S. Schennach, R. Stoch, T. Weber and H. Wickert, *Phys. Lett.* **149B**, 59 (1984).

13. See for instance: F. Iachello and A. Arima, "The interacting boson model", Cambridge Un. Press, 1987, and references therein.

14. A.E.L. Dieperink, *Prog. Part. Nucl. Phys.* **9**, 121 (1983).

15. P. van Isacker, K. Heyde, J. Jolie and A. Sevrin, *Ann. Phys.* **171**, 253 (1986), and references therein.

16. See for instance: H.R. Walet, P.J. Brussard and A.E.L. Dieperink, *Phys. Lett.* **B163**, 4 (1985).

17. E. Lipparini and S. Stringari, *Phys. Lett.* **B130**, 139 (1983).

18. N. Lo Iudice, in "Contemporary topics in nuclear structure physics", R.F. Casten et al. (eds.), World Scientific, Singapore, 1988, pp.541-545;
N. Lo Iudice and A. Richter, *Phys. Lett.* **B228**, 291 (1989).

19. D.R. Bes and R.A. Broglia, *Phys Lett.* **B137**, 121 (1984).

20. T. Suzuki and D.J. Rowe, *Nucl. Phys.* **A289**, 461 (1977).

21. See ref.1, chap.6.

22. L. Zamick, *Phys. Rev.* **C31**, 1955 (1985); **C33**, 691 (1986).

23. See for instance ref.11 for a long, though not exhaustive, list of references.

24. I. Hamamoto and S. Aberg, *Phys. Lett.* **145B**, 163 (1984);
I. Hamamoto and C. Rönström, *Phys. Lett.* **B194**, 6 (1987).

25. See for instance: S. Iwasaki and K. Hara, *Phys. Lett.* **144B**, 9 (1984).

26. K. Sugawara-Tanabe and A. Arima, *Phys. Lett.* **B206**, 573 (1988).

27. F.G. Scholtz, R. Nojarov and A. Faessler, *Phys. Rev. Lett.* **63**, 1356 (1989).

GROUP-THEORY APPROACH TO NUCLEAR PHOTOREACTIONS

Giuseppe Maino

ENEA, Divisione Calcolo, viale Ercolani 8, Bologna, Italy.

Abstract

A simple algebraic model is presented to deal with giant dipole resonances in nuclei. Dynamical symmetries are exploited in cases of practical interest and numerical results for shape-transitional regions are discussed. Applications to photon-induced reactions are shown and calculated cross sections are compared with experimental data. Connections with other nuclear models (both geometrical and algebraic) are briefly outlined.

1. Introduction

From beginning of nuclear-structure studies collective properties have been formulated in terms of geometrical models /1,2/. Therefore, also collective motions of nuclei in the MeV energy region of giant resonances have been described within this framework, soon after the discovery /3/ of large peaks in the nuclear photoabsorption cross sections between 10 and 30 MeV. This behaviour has been shown to be a very peculiar feature of all nuclei from light to heavy-mass ones, from ^{16}O to ^{238}U, for instance. Classical models were developed by Goldhaber and Teller /4/ and Steinwedel and Jensen /5/, essentially based on the assumption that the neutron fluid (incompressible or not according to the latter or former model) vibrates against the proton fluid (see fig.1). The Goldhaber-Teller model predicts a mass dependence of the excitation energy of the giant dipole resonance (GDR) proportional to $A^{-1/6}$, with A mass number, which is experimentally observed in light-mass nuclei. For heavy nuclei the GDR excitation energy is a mixture of Goldhaber-Teller and Steinwedel-Jensen laws (this latter gives an energy $\hbar\omega \simeq 70\ A^{-1/3}$).

More recently /6/, the geometrical model applied to giant resonances have been extended by the Frankfurt group. Its *dynamical collective model* couples giant-resonance degrees of freedom to surface degrees of freedom and, therefore, is able to explain the fragmentation of the GDR in several components.

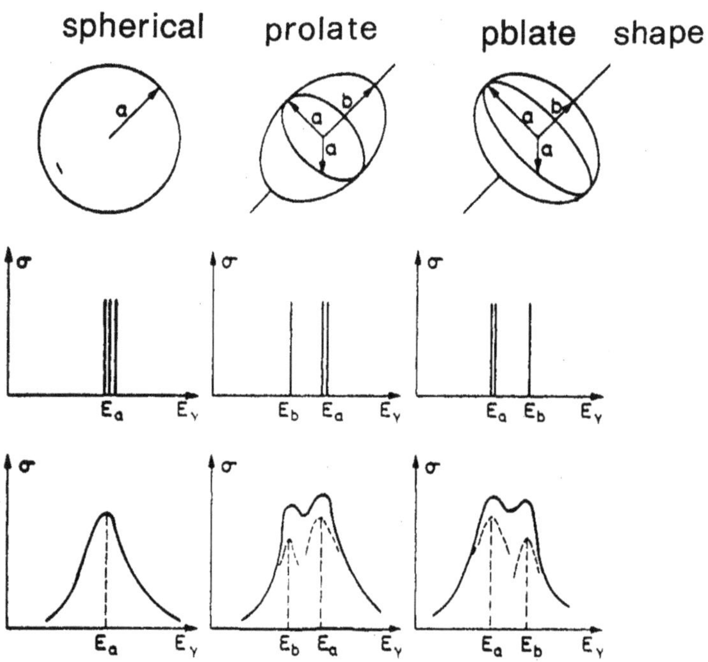

Fig.1 - Geometrical picture of GDR in spherical and deformed nuclei.

The well-known splitting of GDR in two components in deformed nuclei arises because of the ground-state stable deformation; in the case of prolate nuclei, the higher-energy peak is about twice than the lower one since it is doubly degenerate (see fig.1) due to rotations around the symmetry axis.

On the other hand, from a microscopic point of view, the GDR states which carry quantum numbers $J^{\pi} = 1^{-}$ and $T = 1$ in even-even nuclei (respectively, angular momentum and isospin), originate from coherent excitations of many nucleons of $1\hbar\omega$ character. As a consequence, the electromagnetic (dipole) transition probabilities from nuclear ground-state to GDR components have a collective strength one or two orders of magnitude larger than the single-particle transition strength. Moreover, the GDR motion shows a remarkable regularity along all the periodic table, its excitation energy slowly varying with mass number, as above pointed out.

Since $1\hbar\omega$ excitations involve promotions of nucleons across major shells, a complete description of GDR states within the usual shell-model framework is a formidable task if one considers nuclei far away from closed-shell configurations. In fact, the numerical difficulties related to this approach are cumbersome and well above the present capabilities of up-to-date computers.

One has to resort both to geometrical and algebraic models, in order to deal with GDR excitations in non-magic nuclei; in particular, the algebraic approach is well suited since it allows us to emphasize the main characteristics of collective motions and drastically simplify the problem from a computational point of view. Moreover, the algebraic models can be related to the geometrical aspects and interpreted microscopically, thus linking together these two main facets of nuclear structure.

In the last ten years remarkable success has been gained by the Interacting Boson Model (IBM) /7/ of nuclear structure in reproducing many collective properties of nuclei far from closed shells at low energy. Therefore, it deserves careful attention a possible extension of IBM to higher-energy collective degrees of freedom such as giant resonances.

Results obtained for GDR excitations are summarized, together with the relevant formalism, in the following sections.

2. *Formalism*

Low-energy collective states in even-even nuclei are described - in the frame of the interacting boson model /7/ - by means of boson degrees of freedom representing coherent pairs of nucleons with total angular momenta J=0 and 2 (s- and d-boson, respectively). GDR excitation, with intrinsic quantum numbers $J^\pi = 1^-$, is described by a further degree of freedom consisting in a p-boson which obviously carries one unit of angular momentum and negative parity /8,9/. This kind of boson mimics 1p-1h excitation across a major shell rather than collective excitations within the valence shell as s- and d-bosons do.

The algebraic structure of the problem is then given by the usual U(6) symmetry of the IBM /7/, coupled to the U(3) symmetry group of the dipole operators representing the GDR excitation:

$$U(6) \otimes U(3) \supset SU(6) \otimes SU(3).$$

The nuclear states are then classified according to this symmetry chain. The most general Hamiltonian with one- and two-body terms has the following form:

$$\hat{\mathcal{H}} = \hat{\mathcal{H}}_{s,d} + \hat{\mathcal{H}}_{p} + \hat{\mathcal{H}}_{s,d,p}, \tag{1}$$

where $\hat{\mathscr{H}}_{s,d}$ is the usual IBM-1 /7/ Hamiltonian,

$$\hat{\mathscr{H}}_{s,d} = \varepsilon \hat{n}_d + a_0 P^+.P + a_1 \hat{L}.\hat{L} + a_2 \hat{Q}.\hat{Q} + a_3 \hat{T}_3.\hat{T}_3 + a_4 \hat{T}_4.\hat{T}_4, \qquad (2)$$

with

$$\hat{n}_d = d^+.\tilde{d},$$

$$\hat{P} = \frac{1}{2}(\tilde{d}.\tilde{d}) - \frac{1}{2}(\tilde{s}.\tilde{s}),$$

$$\hat{L} = \sqrt{10}\,[d^+ x \tilde{d}]^{(1)},$$

$$\hat{Q} = [d^+ x \tilde{s} + s^+ x \tilde{d}]^{(2)} + \chi\,[d^+ x \tilde{d}]^{(2)},$$

$$\hat{T}_3 = [d^+ x \tilde{d}]^{(3)},$$

$$\hat{T}_4 = [d^+ x \tilde{d}]^{(4)}. \qquad (3)$$

$\hat{\mathscr{H}}_p$ corresponds to the unperturbed p-boson energy:

$$\hat{\mathscr{H}}_p = \varepsilon_p \hat{n}_p, \qquad (4)$$

with $\hat{n}_p$ number of p-bosons, equal to 0 or 1 for low- or high-energy (GDR) levels, respectively.

Finally, $\hat{\mathscr{H}}_{s,d,p}$ describes the coupling between low- and high-lying degrees of freedom and is responsible for the fragmentation of the GDR:

$$\hat{\mathscr{H}}_{s,d,p} = b_0\,[d^+ x \tilde{d}]^{(0)}.[p^+ x \tilde{p}]^{(0)} + b_1\,[d^+ x \tilde{d}]^{(1)}.[p^+ x \tilde{p}]^{(1)} + + b_2\,\hat{Q}.[p^+ x \tilde{p}]^{(2)}. \qquad (5)$$

It is worth recalling that /7/ $\tilde{b}_{jm} = (-1)^{j+m}\, b_{j-m}$.

In eqs.(4,5) ε_p, b_0, b_1 and b_2 are free parameters to be adjusted on the basis of available experimental information, in the present phenomenological approach. In principle, they could be deduced microscopically by a correspondence between IBM and shell model /7/. However, it represents a very difficult task and lies outside the domain of this presentation.

The free p-boson energy, ε_p, is assumed to follow the semiempirical law $\varepsilon_p \simeq 70\,A^{-1/3}$ MeV; the importance of various coupling coefficients,

b_i, $i=0,1$ and 2 , has been extensively discussed in ref./10/. The dominant term is the quadrupole-quadrupole interaction whose strength is mainly responsible for the GDR fragmentation.

In general, Hamiltonian (3-5) must be diagonalized numerically in a basis of states given by

$$|\psi> \;=\; (p^+)^q\,(d^+)^m\,(s^+)^n\;|0>, \tag{6}$$

where $|0>$ is the boson vacuum and $n+m+q=N$, number of bosons; $q=0$ or 1 for low-energy or GDR levels, respectively.

The complexity of the problem is largely reduced with respect to shell-model calculations, even in a suitable truncated basis. Dimensions of the involved matrices in the algebraic IBM approach range from 5-10 up to one hundred.

In addition to the Hamiltonian, one needs a boson dipole operator in order to evaluate the transition probabilities between low-energy states and GDR components. The simplest choice is

$$\hat{D}^{(1)} \;=\; D_0\,[p^+ \times \tilde{p}]^{(1)}, \tag{7}$$

with D_0 adjustable parameter. With these ingredients, it is possible to evaluate excitation energies and transition strengths for GDR states of even-even nuclei far from closed shells.

It is interesting to investigate mass regions where shape transitions are likely to occur, since these transitional isotopes are difficult to study within other models. It is well known that IBM admits three limit symmetries, described by U(5), SU(3) and O(6) group structures corresponding to spherical vibrators, rotors with axial symmetry and nuclei soft with respect to gamma deformations. In figs.2-4 the characteristic splittings of GDR components are shown for schematic Hamiltonians intermediate between these limits.

The transition from a spherical U(5) nucleus to a deformed one, belonging to the SU(3) symmetry, is given by Hamiltonian (1,2) with coefficients ε = 300 keV, $\chi = -\sqrt{7}/2$, ε_p = 14.8 MeV, b_2 = 600 keV and the other parameters put equal to zero. The total boson number, N, is assumed equal to 8 and the quadrupole-quadrupole strength, a_2, decreases from -1.0 keV to -50 keV. Thus, one obtains a fragmentation pattern where several components with similar strengths characteristic of a spherical U(5) nucleus collapse to just two states, the higher being nearly twice than the lower in energy. This situation is typical of a deformed rotor with axial symmetry and roughly corresponds to that predicted by the geometrical model (see fig.1). Some differences betwwen the two model will be discussed later.

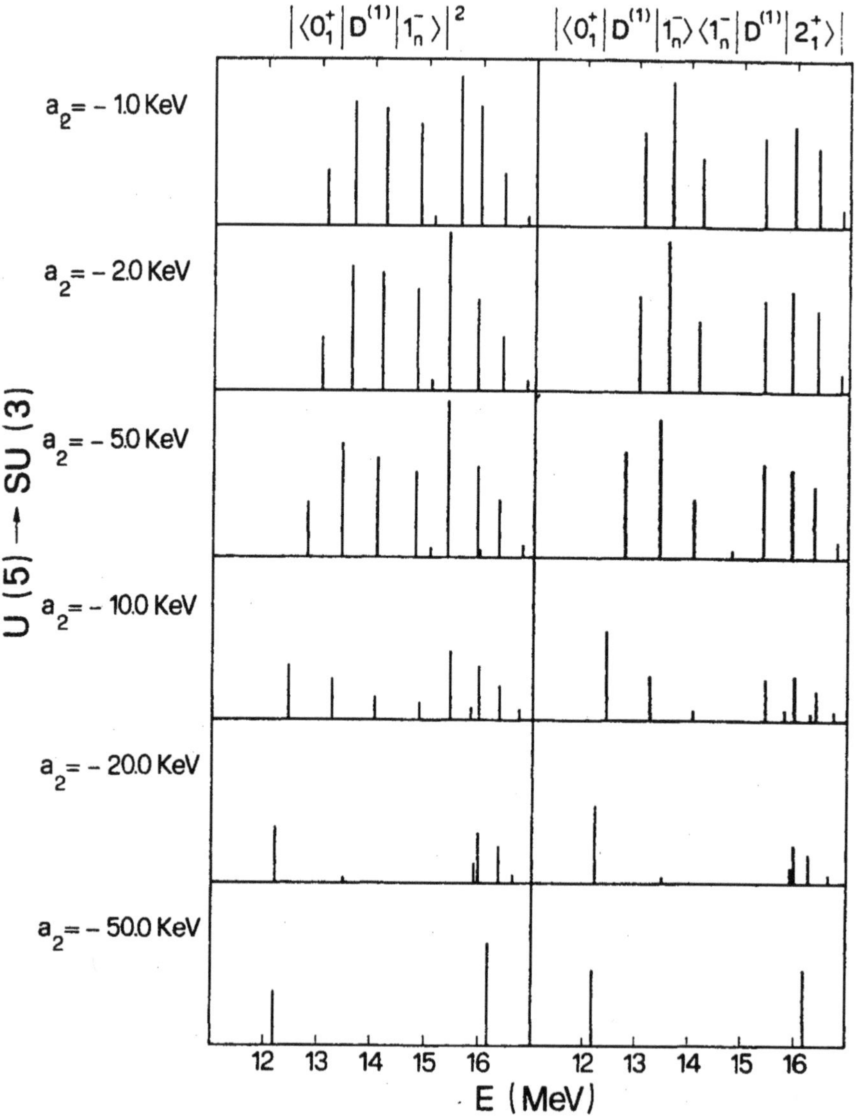

Fig.2 - Energy splittings and dipole strengths of GDR components along a U(5) → SU(3) transitional region. Matrix elements are in arbitrary units.

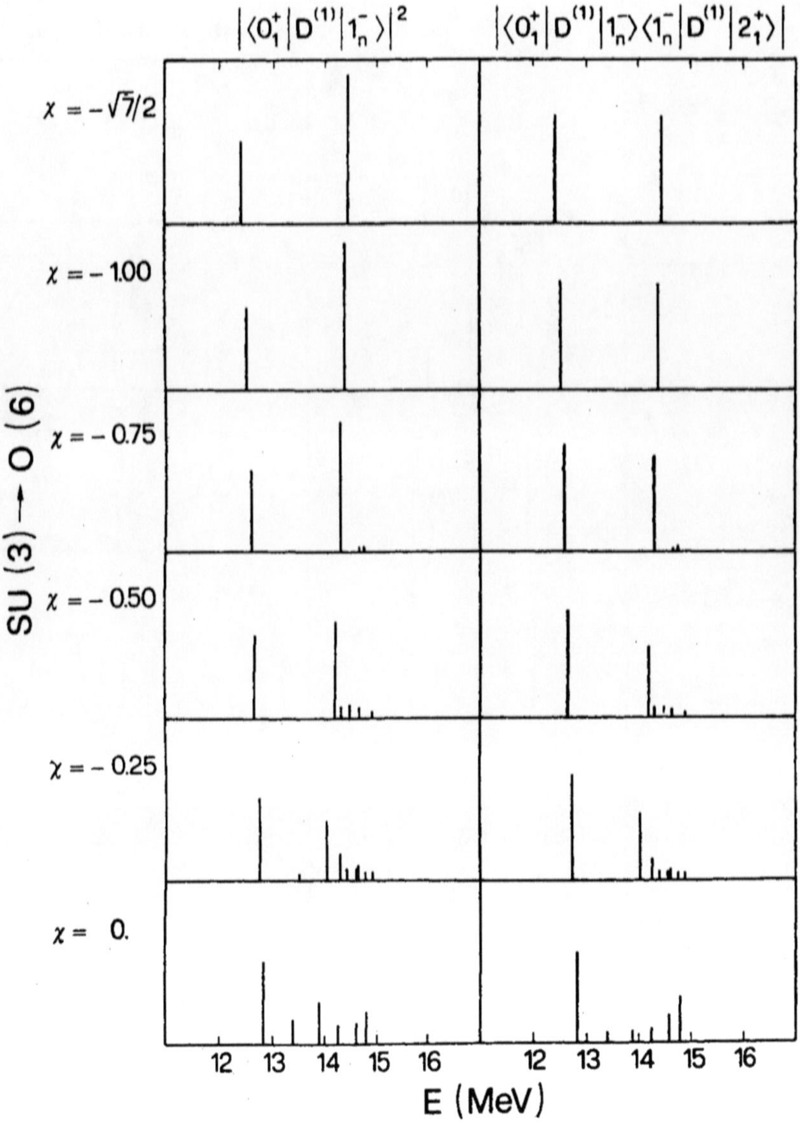

Fig.3 - Energy splittings and dipole strengths of GDR components along a SU(3) → O(6) transitional region. Matrix elements are in arbitrary units.

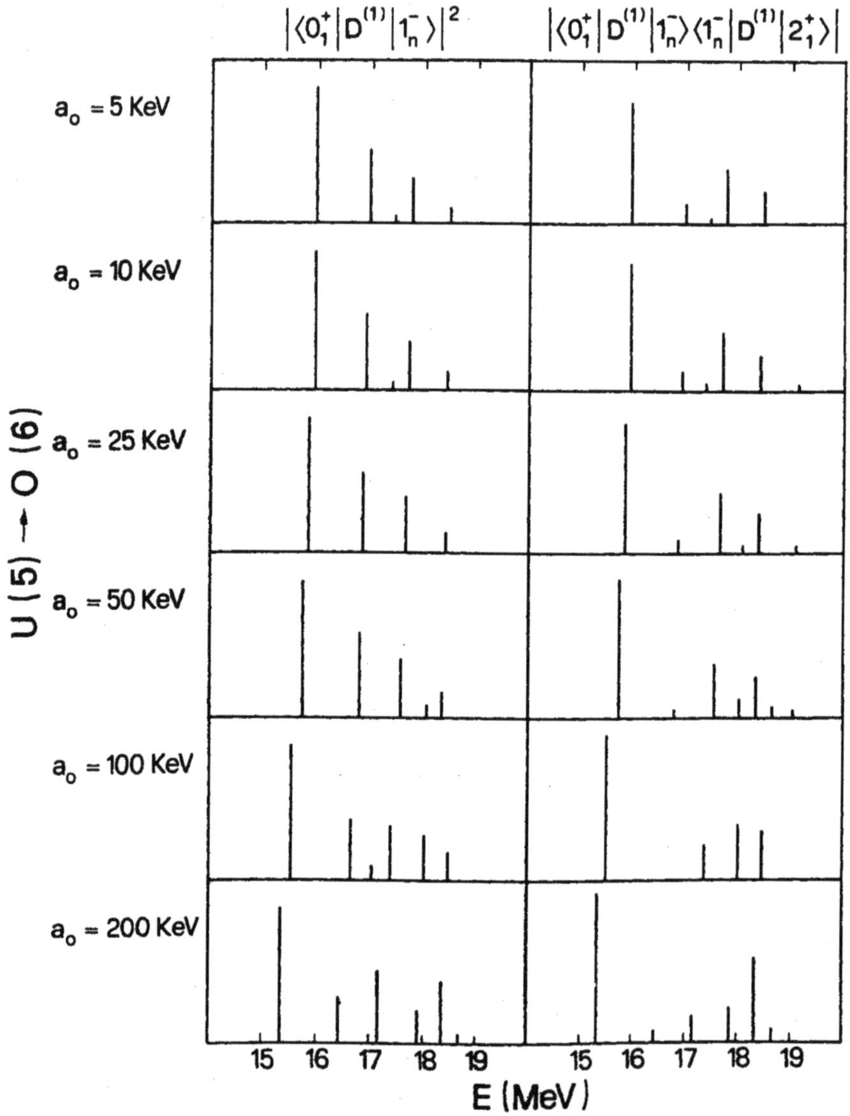

Fig.4 - Energy splittings and dipole strengths of GDR components along a U(5) → O(6) transitional region. Matrix elements are in arbitrary units.

The SU(3) → O(6) shape transition can be recovered by assuming in Hamiltonian (2,3) a_1, a_2, ε_p and b_2 different from zero and constant (the other parameters are taken equal to zero), together with the boson number, N=8. Parameter χ in eq.(3) varies from 0, corresponding to a nucleus with O(6) symmetry, to $-\sqrt{7}/2$ which is the value required by the SU(3) symmetry /7/.

The γ-soft nucleus in the O(6) IBM limit exhibits a typical GDR fragmentation in several states of decreasing strength with increasing excitation energy.

Finally, the U(5) → O(6) case, described by a schematic Hamiltonian with 8 bosons and only ε, ε_p, b_2 different from zero, parameter a_0 changing from 5 kev, near the U(5) limit, to 200 keV (O(6) limit), is presented in fig.4.

In order to compare the IBM predictions about GDR fragmentation and transition probabilities with experimental data, an intrinsic width must be associated with each calculated GDR component. These widths arise from the coupling of GDR states to more complicated nuclear configurations, as 2p-2h doorway states or the continuum. Since these configurations lie outside the IBM model, their effect has to be accounted on a phenomenological "ad hoc" manner. According to ref./2/, the GDR widths are assumed to follow the law $\Gamma(E) \simeq k\,E^{\gamma}$ where the excitation energy, E, is expressed in MeV and k and γ are adjustable parameters to be kept constant for a given mass region.

3. Photon-induced reactions

At this point, it is possible to evaluate photoabsorbtion and scattering cross sections by resorting to usual formalisms /2/. Photon-induced reactions are very convenient tools for the investigation of nuclear structure properties since the theory of electromagnetic interactions is well founded and allows information about nuclear wavefunctions to be univocally determined.

The total photoabsorption cross section in the energy region of GDR excitations, where the long wavelength approximation is applicable, is expressed in terms of nuclear polarizabilities /2/ defined as follows:

$$P_L = \delta_{LL_f} \frac{e^2}{[3(2L_f+1)]^{1/2}} \frac{EE'}{(\hbar c)^2} \sum_{n=1}^{n_{\max}} < L_f^+ || \hat{D}^{(1)} || 1_n^- >$$

$$x \; <1_n^- || \hat{D}^{(1)} || 0_1^+> \left(\frac{1}{E_n + E' + i\Gamma_n/2} + \frac{(-1)^L}{E_n - E - i\Gamma_n/2} \right)$$

$$- \; \delta_{L0} \, \delta_{LL_f} \frac{\sqrt{3}}{2} \frac{(Ze)^2}{Am_N c^2}, \qquad (8)$$

where L is the nuclear angular momentum, L_f (=0 or 2) the final-state angular momentum, E and E' the incident and scattered photon energy, respectively, Γ_n the width of the nth dipole state and n_{max} the number of GDR components. The last term at the right-hand-side of eq.(8) is the Thomson scattering amplitude, where Z and A are the atomic and mass numbers of the target nucleus and m_N the atomic mass unit.

Other contributions to the photon elastic scattering channel, such as Delbrück and atomic Rayleigh amplitudes /11,12/, are negligible for photon energies greater than 10 MeV and sufficiently large scattering angles and, therefore, there will not be described here.

The absorption cross section for unpolarized photons is then given by the optical theorem and results

$$\sigma_{abs} = \frac{4\pi}{\sqrt{3}} \frac{\hbar c}{E} \, \mathrm{Im} P_0. \qquad (9)$$

The differential cross sections for photon scattering at an angle θ are expressed as

$$\frac{d\sigma}{d\Omega}(E, \theta; 0_1^+ \rightarrow L_f^+) = \frac{E'}{E} \, |P_L|^2 \, \delta_{LL_f} g_L(\theta),$$

$$g_0(\theta) = (1 + \cos^2\theta)/6,$$

$$g_2(\theta) = (13 + \cos^2\theta)/12. \qquad (10)$$

Photon absorption and scattering cross sections for isotope chains in shape-transitional regions are calculated in a number of papers and compared with corresponding experimental data. In fig.5 results for Sm isotopes are shown /13/: the transition from a single-humped peak, typical of a spherical nucleus, to a double-humped structure in deformed ^{154}Sm is clearly reproduced by the IBM calculations. The U(5) → SU(3) transition is also discussed in ref./14/, while results relevant to the remaining two transitional cases, namely U(5) → O(6) (Pd isotopes) and SU(3) → O(6) (Os and Pt nuclei), are shown in refs./15/ and /16/, respectively.

In all these cases, where experimental data are available from the literature, the shapes of the total photon absorption cross sections are well reproduced by the theoretical calculations; however, as pointed out in

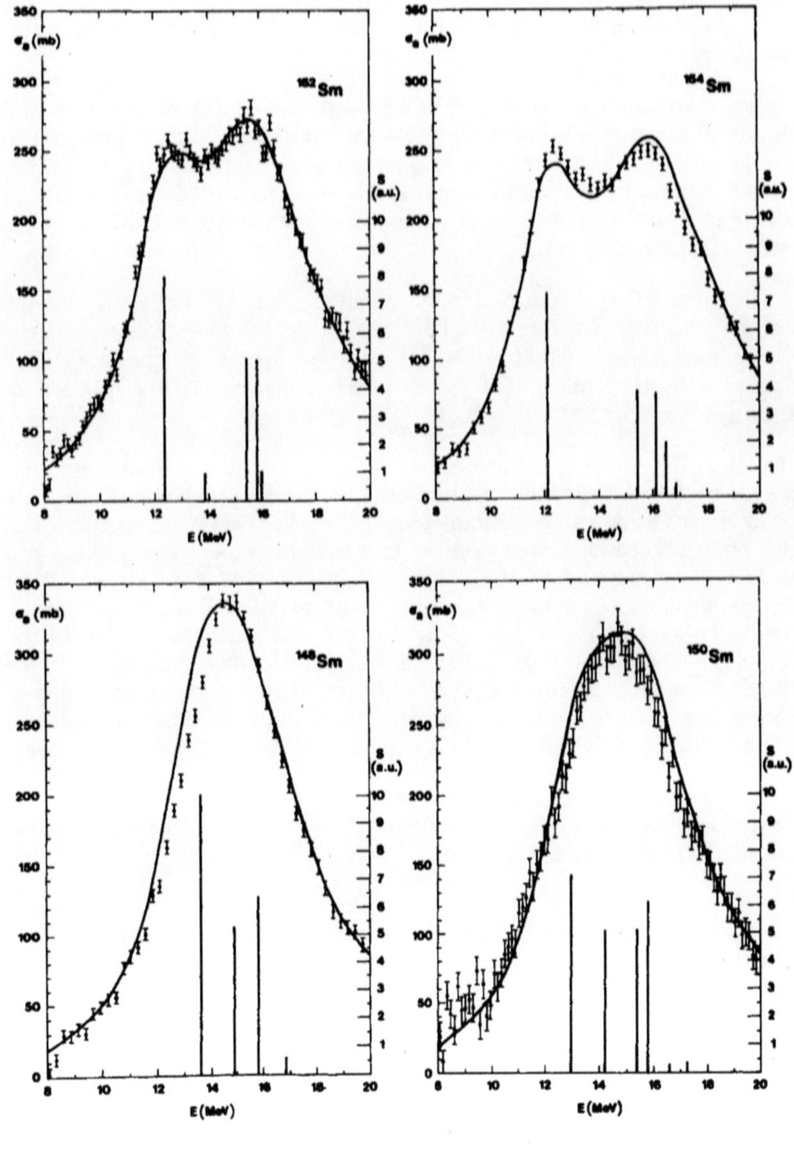

Fig.5 - Experimental and calculated (IBM) total photoabsorption cross sections for Sm isotopes. Experimental data are quoted in ref./13/. The fragmentation of the dipole transition strength, in arbitrary units, is shown at the bottom.

ref./13/, this is *not* a sensitive test of nuclear structure models since the quite large spreading widths, $\Gamma(E)$, tend to wash out any detailed information about nuclear collective motions.

In any case, some differences arise between IBM algebraic approach and the usual geometrical model: for instance, the ratio of dipole strengths to the longitudinal and transverse components in deformed nuclei. The expected value for a classical rotor is 2 /17/, while the IBM result is always less than 2, due to the finite boson number /13,18/, consistently with experimental information.

Moreover, the p-boson model predicts the higher peak in the photoabsorption cross section being larger than the lower one, where the dipole strength is usually confined to a single state. The high-energy peak contains the contribution of different components, slightly splitted, as shown in fig.5 for ^{154}Sm.

On the other hand, photon elastic and inelastic cross sections represent a very useful and sensitive tool in order to investigate nuclear structure properties. The adjustable parameters of the model are fitted to reproduce the photoabsorption cross section and are no more free. Scholtz and Hahne /9/ calculated inelastic Raman scattering to the side bands of ^{168}Er and found remarkable agreement with measurements, while the dynamic collective model was not able to reproduce the experimental findings. Analogous results have been obtained by Nathan /19/ for ^{146}Nd. The geometrical dynamic collective model needs some improvements, discussed in ref./20/, in order to reproduce the observed differential cross sections within the same accuracy as the algebraic IBM model.

Therefore, elastic and inelastic (Raman) scattering cross sections of photons by nuclei represent a stringent test of adopted nuclear models, above all for deformed nuclei /21/.

A careful comparison of dynamic collective model and dipole boson model is carried out in refs./22,23/, making use of the method due to Hatch and Levit /24/ to cast the IBM approach into a geometrical form. The analysis points out the differences between the two models as far as photon inelastic scattering to side bands of deformed nuclei is concerned. Essentially, the dynamic collective model in its usual formulation, without additional terms /20/, strongly couples dipole and surface modes together, thus leading to large inelastic cross section; in the IBM frame, β and γ excited bands belong to different irreducible representations from that of the ground-state rotational band. Therefore, inelastic scattering to side bands is largely suppressed, in agreement with the experimental evidence.

4. GDR on excited states

With the formalism presented in section 2, it is a simple matter to evaluate photoabsorption cross sections on excited states, in addition to that relevant to the ground-state, discussed in the previous section.

The photoabsorption cross section is easily derived from the optical theorem by means of standard techniques and reads as:

$$\sigma_{abs}(E,I_0) = \frac{4\pi e^2}{\hbar c}\,\frac{E^2}{3(2I_0+1)}\sum_{j,k} |<J_k||\hat{D}^{(1)}||I_0>|^2\, x$$

$$\frac{2E_{j,k}\Gamma_{j,k}}{(E_{j,k}^2 - E^2)^2 + \Gamma_{j,k}^2 E^2}, \qquad (11)$$

where E is the incident photon energy, I_0 the spin of excited level on which the GDR is built, J the angular momentum of GDR states ($|I_0-1| \le J \le (I_0+1)$). $E_{j,k}$ and $\Gamma_{j,k}$ are the excitation energy and intrinsic width of each GDR component, respectively.

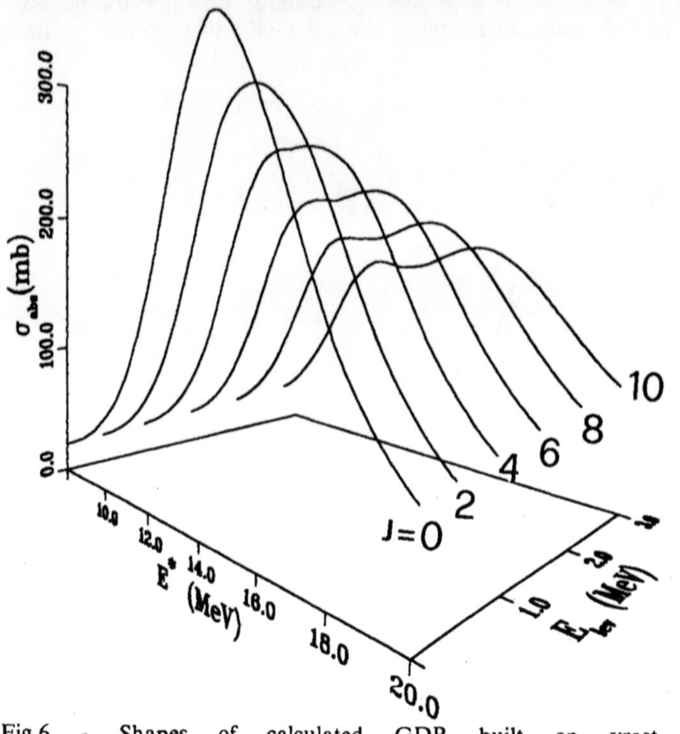

Fig.6 - Shapes of calculated GDR built on yrast levels of ^{150}Sm (0^+, 2^+, 4^+, 6^+, 8^+ and 10^+, respectively).

Assuming that the parameters in Hamiltonian (3-5) do not depend on the initial spin value, the absorption cross sections of figs.6 and 7 result for GDR built on yrast states of ^{150}Sm and ^{154}Sm, respectively. The GDR width slightly increases with increasing spin values and, as for ^{150}Sm, it splits into two components, characteristic of a deformed shape, when $I_0 \geq 8$.

Therefore, the well-known Brink-Axel hypothesis /25/, stating that the energy dependence of the photoabsorption cross section is independent of the detailed structure of the initial level, is valid only as a first approximation. In fact, the GDR mean energy does not seem to depend on the initial excitation energy or spin value, otherwise the total width, as previously outlined.

Unfortunately, no experimental information is available since shapes of GDR built on excited states have been mainly deduced for finite temperature excitations from γ-ray cascades following heavy-ion collisions /26/. Some amount of information for GDR on low-lying discrete levels has been derived from (p, γ) experiments on light-mass nuclei. In this mass region, isospin plays a major role and must be included explicitly in the IBM model, as discussed in section 6. Extension of the dipole boson model to excited states of nuclei in the s-d shell will be object of future works.

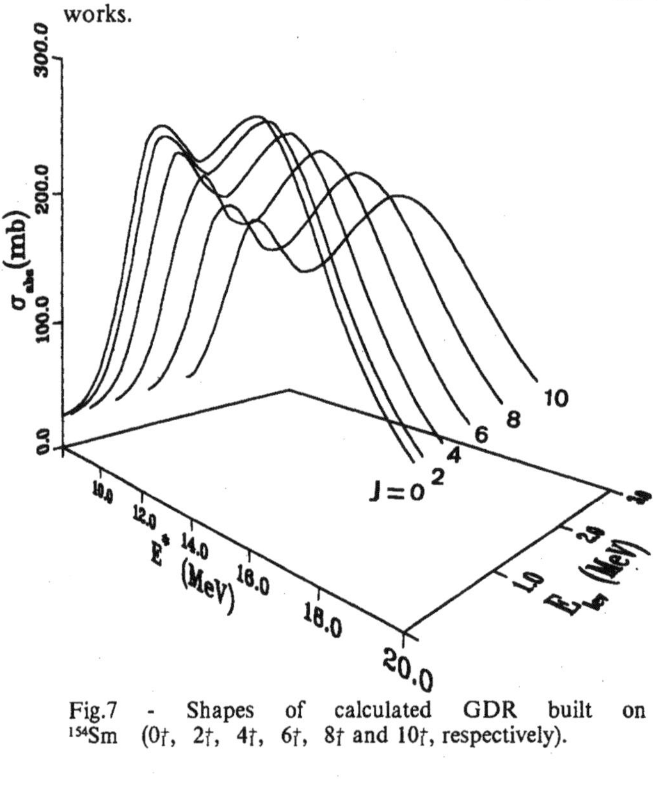

Fig.7 - Shapes of calculated GDR built on yrast levels of ^{154}Sm (0_1^+, 2_1^+, 4_1^+, 6_1^+, 8_1^+ and 10_1^+, respectively).

5. Dynamical symmetries and supersymmetries

The algebraic models of nuclear structure interest because of two other reasons: in some cases, they admit analytical solutions in closed form, when the so-called *dynamical symmetries* /7/ arise. Moreover, due to the underlying group structure, it is easy to add further degrees of freedom such as isospin or single-particle operators (to deal with odd-mass nuclei). These topics will be addressed in this section and the following one.

Since the p-boson operator behaves like a first-rank tensor under SU(3) transformations and belongs to the (1,0) irreducible representation, when low-lying levels are also described by the IBM SU(3) symmetry, corresponding to a stable rotor, the ground-state band belonging to the (2N,0) representation, it is possible to combine together the respective representations thus giving /18/:

$$(2N,0) \otimes (1,0) = (2N+1,0) \oplus (2N-1,1). \qquad (12)$$

The GDR is therefore splitted into two components. Their excitation energies can be easily evaluated /18,27/ if one assumes that all the parameters in Hamiltonian (5) are equal to zero except the quadrupole-quadrupole one, b_2 . The resulting energy splitting is

$$\Delta E = \alpha \sqrt{3}\,(N+\frac{1}{4}), \qquad (13)$$

with α strength of the quadrupole-quadrupole interaction.

The reduced matrix elements of the dipole transitions from the ground-state to the GDR components, eq.(7), are proportional to the Wigner coefficients of the $SU(3) \supset SO(3)$ reduction /18,27/.

With these ingredients it is a simple matter to calculate photon absorption and scattering cross sections by means of formulae (9,10). The relevant results are discussed in refs./10,27/. In fig.8 the total photoabsorption cross section for ^{232}Th is shown, as an example, obtained on the assumption of exact SU(3) symmetry.

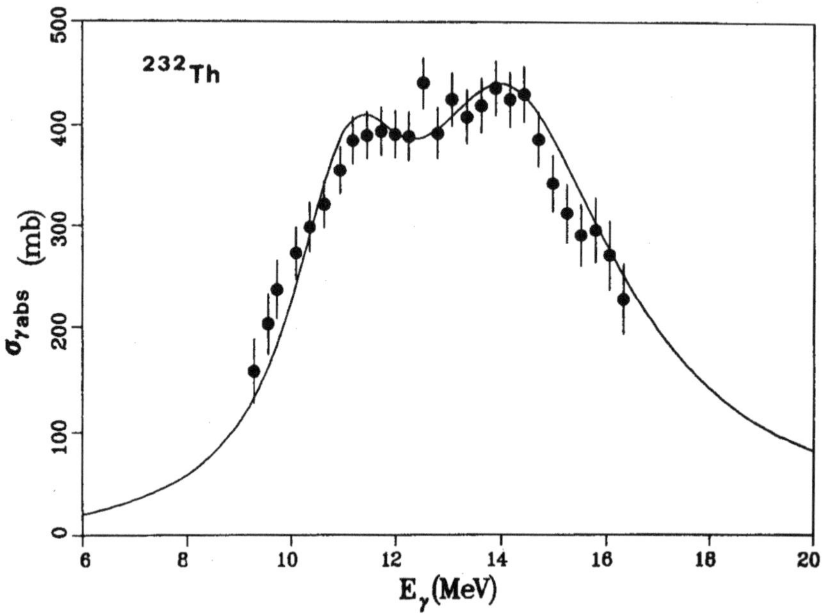

Fig.8 - Experimental and calculated (IBM) total photoabsorption cross section of ^{232}Th in the exact SU(3) symmetry. Experimental data are taken from: A.Veyssiere et al., Nucl. Phys. A 199 (1973) 45.

Moreover, photon elastic and inelastic scattering cross sections are presented in fig.9 where both analytical SU(3) results and realistic calculations for ^{232}Th are compared with experimental data.

A further dynamical symmetry arises for triaxial nuclei. Stable triaxial deformations can not be accounted for in the IBM-1 approximation /7/, where no distinction is made between neutron and proton degrees of freedom. Triaxial shapes are allowed in the more refined IBM-2 version /7/ which takes explicitly into account neutron and proton bosons. Assuming a different kind of deformation for the neutron and proton fluids, i.e. $\chi_\nu = -\chi_\pi = -\sqrt{7}/2$ in the quadrupole operator, $\hat{Q}$, of eq.(3), and considering the group chain /7/:

$$U_\pi(6) \otimes U_\nu(6) \supset \overline{SU_\pi(3)} \otimes SU_\nu(3) \supset SU_{\pi\nu}(3)^* \supset SO_{\pi\nu}(3) \supset SO_{\pi\nu}(2), \tag{14}$$

with labels π and ν referring to proton and neutron bosons, respectively, a dynamical symmetry may occur that corresponds to deformed nuclei with triaxial shape.

The SU(3) representation to which the ground-state belongs is $(2N_\nu,\ 2N_\pi)$ with $N_\nu \geq N_\pi$. The appropriate product representations for GDR states are /18,27/

$$(2N_\nu, 2N_\pi) \otimes (1,0) = (2N_\nu+1, 2N_\pi) \oplus (2N_\nu-1, 2N_\pi+1) \oplus (2N_\nu, 2N_\pi-1). \tag{15}$$

The GDR is therefore splitted into three components, not yet resolved in any measured photoabsorption cross section for medium- and heavy-mass nuclei. Excitation energies and dipole transition strengths of these states are given in refs./18,27/.

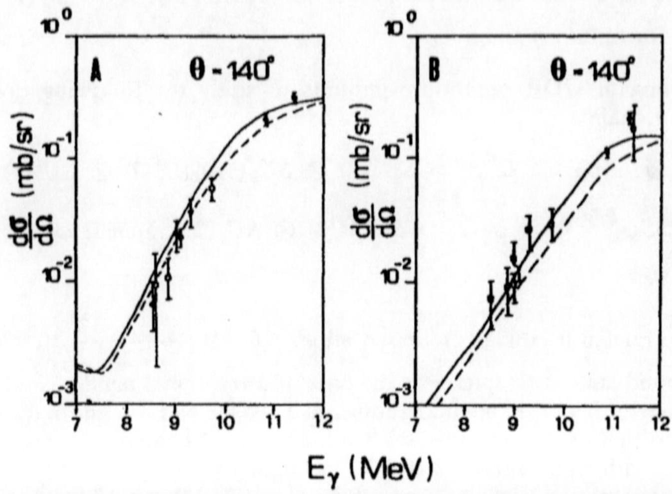

Fig.9 - Experimental and calculated (IBM) cross sections of photon (A) elastic and (B) inelastic scattering to the 2_1^+ of ^{232}Th at an angle $\theta = 140°$. Solid line: realistic calculations with broken SU(3) symmetry; dashed line: analytical results on the assumption of exact SU(3) symmetry. Experimental data are quoted in ref./21/.

We have dealt with even-even nuclei so far; however, the GDR p-boson model can be also applied to odd-mass nuclei by taking into account single-particle degrees of freedom in addition to boson ones.

Low-lying levels of odd-mass nuclei can be satisfactorily described within the framework of the Interacting Boson-Fermion model /28/, where boson and fermion degrees of freedom are included. The latter describes the motion of unpaired nucleons. Within this context, dynamic Bose-Fermi symmetries are of particular interest both from a theoretical and applied point of view; they occur when a usual IBM symmetry valid for the even-even boson core and one subgroup of U(m), symmetry group of single-particle states with

$$m = \sum_{j=1}^{n}(2j+1)$$

and n number of single-particle active shells, can be combined together in a common boson-fermion group chain and the relevant Hamiltonian expressed in terms of Casimir operators of the chain subgroups.

Possible Bose-Fermi symmetries associated with IBM SU(3) boson symmetry have been extensively investigated /28,29/, based on the splitting of the angular momentum, j, of the unpaired nucleon into a pseudo-orbital, k, and a pseudo-spin part, s. For instance, when j = 1/2 and 3/2, k = 1 sits in the three-dimensional representation, [1], of $SU^F(3)$ (F label stands for fermion) and $s = \frac{1}{2}$.

Coupling of GDR excitation amounts to study the following decomposition chain:

$$U^B(6) \otimes U^F(m) \otimes U^P(3) \supset SU^B(3) \otimes SU_k^F(3) \otimes U_s^F(2) \otimes SU^P(3) \supset$$
$$\supset SU^{BFp}(3) \otimes U_s^F(2) \supset SO^{BFp}(3) \otimes SU_s^F(2) \supset Spin(3) \supset Spin(2), \tag{16}$$

and the related irreducible representations /30/. If $j = \frac{1}{2}, \frac{3}{2}$, m = 6 and the ground-state with spin $\frac{1}{2}$ is the base of a rotational band. The GDR states have, therefore, angular momenta J = 1/2 and 3/2 and parity opposite to that of the ground-state band. Five GDR components arise, two of them with spin values 1/2 and 3/2 belonging to the (2N + 2,0) irreducible representation, the remaining three (J = 1/2, 3/2 and 3/2) to (2N,1).

The energy splitting can be evaluated analytically /30/ by considering the following Hamiltonian:

$$\hat{\mathscr{H}} = \hat{\mathscr{H}}(s,d,a_j) + \varepsilon_p \hat{n}_p - \alpha \frac{\sqrt{3}}{2} \hat{Q}^{BF}.[p^+ x \tilde{p}]^{(2)}, \tag{17}$$

where a_j^+ (a_j) creates (annihilates) one nucleon in the j-shell, $\hat{\mathscr{H}}(s,d,a_j)$ is the usual IBFM Hamiltonian /28/ describing the low-energy spectrum of odd-mass nuclei and the boson-fermion quadrupole operator is defined as

$$\hat{Q}^{BF} = (d^+ x\tilde{s} + s^+ x\tilde{d})^{(2)} - \frac{\sqrt{7}}{2}(d^+ x\tilde{d})^{(2)}$$
$$\mp \frac{1}{\sqrt{2}}[(a_{1/2}^+ \, x \, \tilde{a}_{3/2})^{(2)} - (a_{3/2}^+ \, x \, \tilde{a}_{1/2})^{(2)} - (a_{3/2}^+ \, x \, \tilde{a}_{3/2})^{(2)}], \tag{18}$$

with the sign - or + depending on the particle- or hole-coupling character /30/. Once expressed Hamiltonian (17) in terms of quadratic Casimir operators of SU(3), an approximate diagonalization yields the energy splitting between the two GDR components with spin 1/2 :

$$\Delta E = \alpha \sqrt{3}\,(N + \frac{5}{4}), \qquad (19)$$

while corresponding results for spin-3/2 components have been obtained in ref./30/. Since the quadrupole-quadrupole interaction, α, has the same strength for neighbouring nuclei, by comparison of eq.(19) with eq.(13) it results a larger GDR splitting in odd-even isotopes than in even-even adjacent ones.

6. The role of isospin

For light-mass nuclei, where neutrons and protons occupy the same major shells, the possibility of correlated neutron-proton pairs is no more negligible. Therefore, neutron-proton bosons must be considered in addition to the usual proton and neutron ones of IBM-2.

The relevant model, named IBM-3 /31/, assumes three kind of s and d boson excitations, consisting of neutron-neutron, neutron-proton and proton-proton correlated pairs. They belong to an isospin triplet with third component equal to +1, 0 and -1, respectively. If one adds to each boson an intrinsic 'spin' component in addition to the 'orbital' one (equal to 0 and 2), a further extension of the model occurs (IBM-4). Both these versions have been extensively studied /31/ and applied to low-energy spectroscopy of light-mass nuclei, in the s-d shell for instance. Due to the existence of neutron-proton bosons it is also possible describe low-energy spectra of odd-odd nuclei within the IBM-3 or -4 framework.

Since the GDR is an isovector excitation, it is possible to combine together IBM-3 (or -4) and dipole-boson model. This implies that the following group decomposition chains in the isospin space have to be properly added to the boson groups considered in the previous sections:

$$SU_T^{sd}(3) \otimes SU_T^{p}(3) \supset \begin{cases} SU_T^{sdp}(3) \supset SU_T^{sdp}(2), \\ SU_T^{sd}(2) \otimes SU_T^{p}(2) \supset SU_T^{sdp}(2), \end{cases} \qquad (20)$$

where index T labels isospin groups.

The former decomposition chain in eq.(20) corresponds to a strong isospin coupling, while the latter is valid within a weak coupling scheme. Both symmetries have been studied in refs./32,33/; the relevant

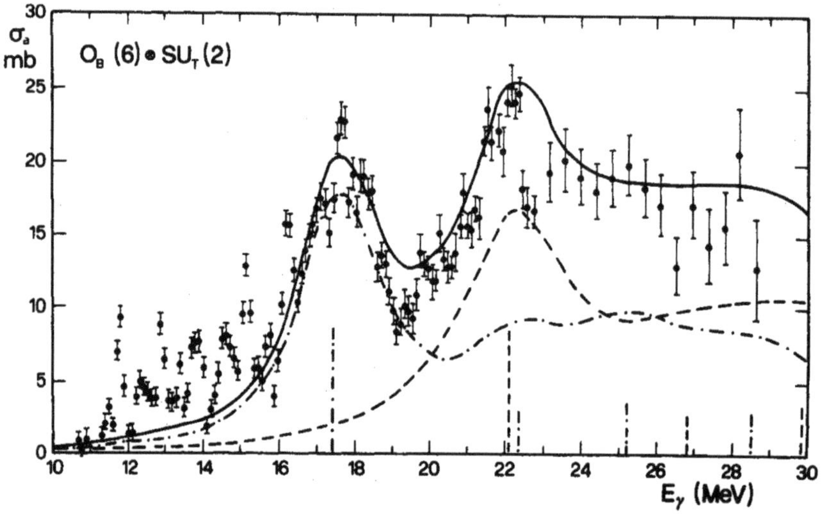

Fig.10 - Experimental and calculated (IBM with broken $O^B(6)$ symmetry) total photoabsorption cross section of ^{26}Mg. Dotted-dashed line: T = 1 component; dashed line: T = 2 components; solid line: sum of tthe two isospin contributions. Dipole strengths in arbitrary units are shown at the bottom. Experimental data are quoted in ref./32/.

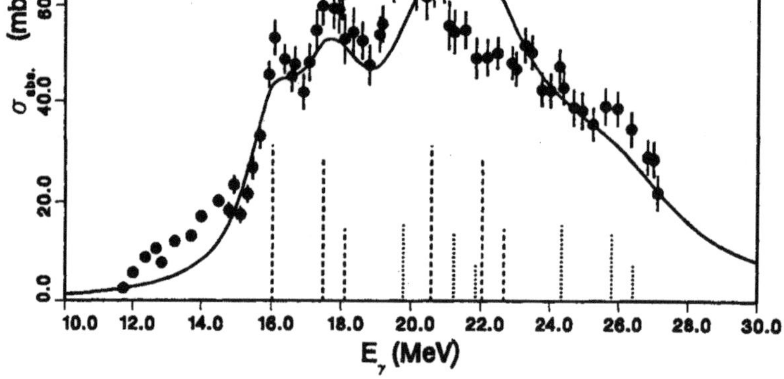

Fig.11 - Experimental and calculated ($SU^B(3) \otimes SU^T(2)$) total photoabsorption cross section of ^{48}Ti . Bars at the bottom represent the fragmentation of the dipole strength, in arbitrary units; dashed line: T = 2 component; dotted line: T = 3 component.

Hamiltonian must be consequently generalized by including isospin degree of freedom and a further splitting of GDR components arise because of different isospin values.

In actual cases, the weak-coupling limit gives satisfactory results as far as comparisons between measured and calculated photoabsorption cross sections are concerned. Figs. 10 and 11 show some results obtained for ^{26}Mg and ^{48}Ti, belonging to the s-d and f-p shells, respectively.

7. Multiplets of giant resonances

It is worth mentioning that giant resonances other than the isovector dipole one can be described within the IBM framework by means of boson degrees of freedom. For instance, giant isoscalar monopole and quadrupole resonances are simulated in a phenomenological algebraic approach /34/ by the introduction of s′ and d′ boson excitations with angular momenta and parity $J^\pi = 0^+$ and 2^+ , respectively, and isospin T=0.

Therefore, due to the common underlying algebraic structure, it is possible to classify many giant resonances into multiplets of a given symmetry group, analogously to low-energy collective levels. The Wigner supermultiplet theory, based on SU(4) symmetry, has been applied to the particle-hole model for the description of inelastic electron scattering form factors /35/, when monopole, dipole, dipole spin-flip and quadrupole modes are excited.

Moreover, the geometrical Goldhaber-Teller model has been generalized likewise /36/, allowing for surface oscillating modes and spin and isospin degrees of freedom. This model too has been used in the investigation of electron scattering experiments.

Some years ago, a unified description of nuclear collective modes, including giant resonances, has been proposed within a pure algebraic scheme /37/, considering isoscalar s bosons and isovector p bosons. Giant resonances can then be described according to the representations of boson $U_L(4) \otimes U_T(4)$ group, where indexes L and T refer to boson angular momentum and isospin, respectively.

A possible hydrogen-like structure for giant resonances based on the symmetry-group chain

$$U_S(4) \otimes U_T(4) \supset SO_S(4) \otimes SO_T(4) \supset SO_S(3) \otimes SO_T(3), \qquad (21)$$

with S and T labelling spin and isospin degrees of freedom, has been recently proposed /38/ and compared with existing experimental information.

Considering the IBM-4 formalism with boson, spin and isospin degrees of freedom /7,31/, an extended version of the algebraic model for giant resonances can be developed which encompasses both Wigner supermultiplet /35,36/ and hydrogen-like /38/ models. In fact, a group structure based on low-lying s and d bosons and high-lying s′ and p bosons such as

$$U_B^{s,d}(6) \otimes U_{ST}^{s,d}(6) \otimes U_B^{s',p}(4) \otimes U_S^{s',p}(4) \otimes U_T^{s',p}(4), \qquad (22)$$

generalizes the above-mentioned models which can be recovered as particular subgroups of all the possible decomposition chains.

8. Conclusions

The main features of the algebraic approach to giant resonances within the IBM framework have been briefly reviewed; this model can provide us with a simple yet reliable tool to investigate nuclear structure properties in the energy region above 10 MeV.

Moreover, the p-boson model can be easily related to a geometrical interpretation. The obvious drawback is that the present algebraic description of giant resonances is only phenomenological and needs some adjustable parameters, to be determined on the basis of available experimental information. A major improvement would be the establishment of a correspondence between the IBM plus p boson and the shell model, such as that developed for IBM-2, in order to deduce microscopically the phenomenological parameters.

On this subject, a recent extension of the symplectic model /39/ has been proposed by Quesne /40,41/, following a suggestion by Rowe and Iachello /18/, including the electric dipole operators among the usual generators of this model /39/. The resulting group structure is given by the following semidirect product

$$W(3) \otimes Sp(6, \mathcal{R}), \qquad (23)$$

where W(3) is the Weyl-Heisenberg group generated by three pairs of boson creation and annihilation operators /40/, while $Sp(6, \mathcal{R})$ is the usual group-structure of the symplectic model. Since this latter is a fully microscopic description of nuclear collective motions, a further analysis of the $wsp(6, \mathcal{R})$ algebra and its connection with the dipole-boson IBM approximation could provide some insight into a microscopic foundation of algebraic models for GDR.

Acknowledgments

A great deal of the work here described has been done in collaboration with A.Ventura and L.Zuffi. Useful discussions with them and F.Iachello are gratefully acknowledged.

References

/1/ A.Bohr and B.R.Mottelson, "Nuclear Structure", vol.II, Reading Massachussetts (1975).

/2/ J.M.Eisenberg and W.Greiner, "Nuclear Theory", vol.1, Amsterdam (1978).

/3/ G.C.Baldwin and G.S.Klaiber, Phys. Rev. 73 (1948) 1156.

/4/ M.Goldhaber and E.Teller, Phys. Rev. 74 (1948) 1046.

/5/ H.Steinwedel, J.H.D.Jensen and P.Jensen, 79 (1950) 1019.

/6/ H.Arenhövel, M.Danos and W.Greiner, Phys. Rev. 157 (1967) 1109.

/7/ F.Iachello and A.Arima, "The Interacting Boson Model", Cambridge (1987).

/8/ I.Morrison and J.Weise, J. Phys. G 8 (1982) 687.

/9/ F.G.Scholtz and F.J.W.Hahne, Phys. Lett. 123*B* (1983) 147.

/10/ G.Maino, A.Ventura and L.Zuffi, in "Perspectives on Theoretical Nuclear Physics", vol.2, L.Bracci et al., eds., Pisa (1988), p.222.

/11/ S.Turrini, G.Maino and A.Ventura, Phys. Rev. C 39 (1989) 824, and references quoted therein.

/12/ R.Moreh, in "Intermediate Energy Nuclear Physics", R.Bergere, S.Costa and C.Schaerf, eds., Singapore (1982) p.1.

/13/ G.Maino, A.Ventura, L.Zuffi and F.Iachello, Phys. Rev. C 30 (1984) 2101.

/14/ F.G.Scholtz, Phys. Lett. 151*B* (1985) 87.

/15/ A.Ventura, G.Maino, P.Van Isacker and L.Zuffi, in "Nuclear Structure, Reactions and Symmetries", R.A.Meyer and V.Paar, eds., Singapore (1986), vol.II, p.957.

/16/ G.Maino, A.Ventura, P.Van Isacker and L.Zuffi, Phys. Rev. C 33 (1986) 1089.

/17/ M.Danos, Nucl. Phys. 5 (1958) 23; K.Okamoto, Phys. Rev. 110 (1958) 143.

/18/ D.J.Rowe and F.Iachello, Phys. Lett. 130*B* (1983) 231.

/19/ A.M.Nathan, in "Nuclear Structure, Reactions and Symmetries", R.A.Meyer and V.Paar, eds., Singapore (1986), vol.I, p.385.

/20/ A.M.Nathan, Phys. Rev. C 38 (1988) 92.

/21/ G.Maino, A.Ventura, L.Zuffi and F.Iachello, Phys. Lett. 152*B* (1985) 17.

/22/ F.G.Scholtz and F.J.W.Hahne, Phys. Rev. C 34 (1986) 693.

/23/ F.G.Scholtz and F.J.W.Hahne, Nucl. Phys. A 471 (1987) 545.

/24/ R.L.Hatch and S.Levit, Phys. Rev. C 25 (1982) 614.

/25/ D.M.Brink, Thesis, Oxford (1955); P.Axel, Phys. Rev. 126 (1962) 671.

/26/ See, for instance, Nucl. Phys. A 482 (1988) 1c and following.

/27/ L.Zuffi, P.Van Isacker, G.Maino and A.Ventura, Nucl. Instr. Meth. Phys. Res. A 255 (1987) 46.

/28/ F.Iachello and P.Van Isacker, "The Interacting Boson-Fermion Model", Cambridge (in press).

/29/ R.Bijker and V.K.B.Kota, Phys. Rev. C 37 (1988) 2149; Ann. Phys. (N.Y.) 187 (1988) 148.

/30/ G.Maino, Phys. Rev. C 40 (1989) 988.

/31/ J.P.Elliott, Rep. Progr. Phys. 48 (1985) 171, and references quoted therein.

/32/ G.Maino, A.Ventura and L.Zuffi, Phys. Rev. C 37 (1988) 1379.

/33/ E.Bortolani and G.Maino, submitted to Phys. Rev. C.

/34/ G.Maino, A.Ventura, P.Van Isacker and L.Zuffi, Europhys. Lett. 2 (1986) 345.

/35/ T.W.Donnelly and G.E.Walker, Ann. Phys. (N.Y.) 60 (1970) 209.

/36/ T.W.Donnelly and N.Lo Iudice, Nucl. Phys. A 179 (1972) 569.

/37/ Hsi-Tseng Chen, Phys. Rev. Lett. 47 (1981) 1382.

/38/ Hsi-Tseng Chen and F.Krmpotic, Phys. Rev. C 32 (1985) 1457.

/39/ D.J.Rowe, Rep. Progr. Phys. 48 (1985) 1419.

/40/ C.Quesne, Phys. Lett. 188*B* (1987) 1.

/41/ C.Quesne, Ann. Phys. (N.Y.) 185 (1988) 46.

ORDERED AND CHAOTIC MOTIONS IN ATOMIC NUCLEI (*)

V.R. Manfredi

Dipartimento di Fisica dell'Università di Padova
INFN, Sezione di Padova
via Marzolo, 8; 35131 Padova - I

ABSTRACT

The co–existence of regular and chaotic motions in atomic nuclei is briefly discussed and the three–level Lipkin, Meshkov, Glick model, whose classical analog shows a smooth transition from regular to chaotic regime, is successively introduced.

1. – INTRODUCTION

An interesting aspect of nuclear dynamics is the co–existence, in atomic nuclei, of "ordered" and "chaotic" motions [1)]. The ordered motion is described by a variety of models, such as the shell–model, the collective model and their various extensions. Roughly speaking, regular nuclear motion can be understood in terms of the mean–field approximation and of the oscillations about the mean field which give rise to collective excitations [2)].

But another aspect of nuclear dynamics is also present both in nuclear spectroscopy and in nuclear reactions: the aspect of irregularity, or of chaos. In particular, as discussed in Section 2, an excellent agreement has been obtained between the level density fluctuations of the Nuclear Data Ensemble (NDE) and the previsions of a Random Matrix Model: the Gaussian Orthogonal Ensemble (GOE) [3),4)].

The co–existence in atomic nuclei of opposite aspects of nuclear dynamics (ordered and chaotic) lead naturally to put the question: is it possible to study the transition from regular to chaotic regime by a simple model?

In this spirit a model has recently been proposed by Meredith & Koonin: the three–level Lipkin, Meshkov, Glick Model (see section 3).

(*) Work supported in part by Ministero della Pubblica Istruzione

2. – LEVEL SPACING DISTRIBUTION

In this section we shall follow very closely the approach of ref. [1]. At excitation energies of several MeV, nuclei display statistical properties. This idea, proposed by N. Bohr in 1936 [5], dominated the interpretation of the nuclear reaction data until the advent of the independent particle model and the discovery of the direct reaction mechanism. Thereafter, the interest in statistical nuclear properties was kept alive by theoretical work on random matrices based on the Dyson and Wigner papers [6,7].

Interest resurged with the discovery of Ericson's fluctuations [8], which showed that the compound nucleus cross section behaves like a random variable. Nevertheless, early comparisons between theory and experiment were inconclusive [3], because there were no data of suitable quality. Only at the end of the seventies were high quality data obtained by neutron resonance spectroscopy and from high resolution proton scattering experiments, resulting from a collective effort performed mainly at Columbia University [9] and TUNL [10].

These data allowed Haq, Pandey and Bohigas [11,12] to establish statistically significant agreement between measured level spacing fluctuations and Wigner's random matrix theory model.

The authors of references [11,12], scaling each sequence to the same average level spacing and putting together all sequences, constructed the so–called "Nuclear Data Ensemble" (NDE) which comprises 1726 level spacings. In order to compare the experimental results with the prediction of GOE (*), two different statistics (**) have been used: the nearest neighbour spacing distribution $P(s)$ and the Δ_3 of Dyson and Metha [4].

(*) The Gaussian Orthogonal Ensemble (GOE) consists of an ensemble of real symmetric matrices whose elements are independent random variables. The name of the ensemble derives from the Gaussian form of the probability density $P(H)$ and from the orthogonal invariance of dH, where dH is the measure in the space of matrices. For deeper discussion, see reference [13].

(**) *"A statistic is a quantity which can be calculated from an observed sequence of levels alone, without other information and whose average value and variance are known from the theoretical model. A suitable statistic is one which is sensitive for the property to be compared or distinguished and is*

The first statistic measures the probability that two neighbouring eigenvalues are a distance s apart. For this statistic the Poisson ensemble average is:

$$P(s) = e^{-s}, \tag{1}$$

which gives a large probability for the occurrence of near degeneracies, while the GOE average is closely approximated by the Wigner distribution:

$$P(s) = \frac{\pi}{2} s \exp\left[-\frac{\pi}{4} s^2\right], \tag{2}$$

which gives level repulsion.

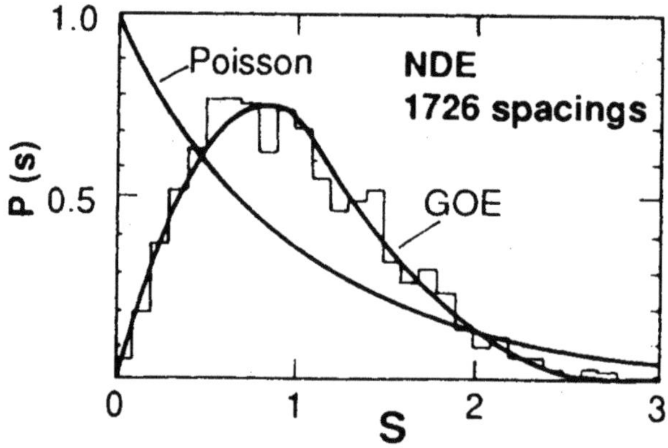

Figure 1. – Nearest–neighbour spacing distribution for the Nuclear Data Ensemble (NDE); adapted from reference 12.

The second statistic (Δ_3) is defined, for a fixed interval $(-L, L)$, as the least–square deviation of the staircase function $N(E)$ from the best straight line fitting it:

---

insensitive for other details" [4].

$$\Delta_3 = \frac{1}{2L} \min_{A,B} \int_{-L}^{L} [N(E) - AE - B]^2 \, dE, \tag{3}$$

where $N(E)$ is the number of levels between E and zero for positive energy, between $-E$ and zero for negative energy. The statistic Δ_3 provides a measure of the degree of rigidity of the spectrum: for a given interval L, the smaller Δ_3, the stronger the rigidity, signifying the long-range correlations between levels [14)15)] .

The agreement between experimental results and GOE predictions is astonishingly good and also surprising since the GOE contains no adjustable parameters (see Figure 1 and Figure 2).

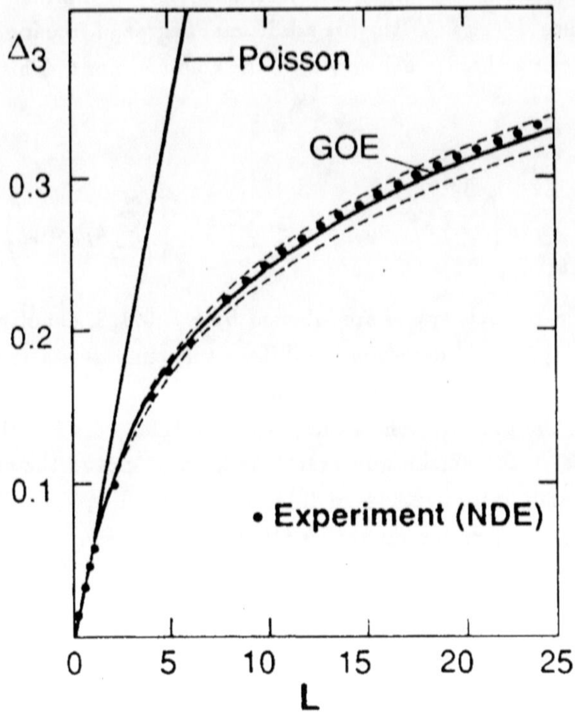

Figure 2. – Δ_3 statistic for the Nuclear Data Ensemble (NDE); adapted from reference 12.

3. – THE THREE LEVEL LIPKIN–MESHKOV–GLICK MODEL

3.1. – *The Model*

As discussed in the introduction, the two–level Lipkin–Meshkov–Glick (LMG) model [16] has recently been extended by Meredith & Koonin [17] in order to study the transition from ordered to chaotic motion. In fact, as discussed in references [17),18] the two–level LMG model has only one degree of freedom (e.g. the number of particles in the upper level) and, therefore, the classic analog cannot exhibit chaotic behaviour at any value of the coupling constant [(*)]. The generalization of the LMG model, proposed in references [17),18] describes a system of M particles in three M–fold degenerate levels, which has two degrees of freedom (e.g. the number of particles in the two highest levels). With this additional degree of freedom, comes the possibility of seeing chaotic behaviour in the classical limit. This three–level Hamiltonian is:

$$H = \sum_{k=0}^{2} \epsilon_k \left(\sum_{m=1}^{M} a_{km}^{+} a_{km} \right) - \frac{1}{2} \sum_{k,\ell=0}^{2} V_{k\ell} \left(\sum_{m=1}^{M} a_{km}^{+} a_{\ell m} \right)^2 , \tag{4}$$

where the single–particle states are labelled by $k = 0, 1, 2$, denoting the three levels and $m = 1, 2, \ldots M$ denoting the different degenerate states within each level. $V_{k\ell}$ is the strength of the interaction.

Following the prescription of the authors of references [17),18] we assume $\epsilon_2 = -\epsilon_0 \equiv \epsilon$, $\epsilon_1 = 0$, vanishing interactions for particles in the same orbital and equal interactions for particles in different orbitals [i.e., $V_{k\ell} \equiv V\,(1 - \delta_{k\ell})$]. Defining the nine bi–fermion operators [17),18]

$$G_{k\ell} \equiv \sum_{m=1}^{M} a_{km}^{+} a_{\ell m} \tag{5}$$

[(*)] For an exhaustive discussion of the chaotic motion in classical mechanics, see reference [19].

the Hamiltonian (4) can be written in terms of these operators only

$$H = \sum_{k=0}^{2} \epsilon_k G_{kk} - \frac{1}{2} \sum_{k,\ell=0}^{2} V\,(1-\delta_{k\ell})\,G_{k\ell}^2. \tag{6}$$

The non interacting ground state has all particles in the ground orbital, and is represented by $\mid 00 >$. The other basis states are written using symmetric raising operators

$$\mid bc >= C\,(b,c)\,G_{10}^b G_{20}^c \mid 00 >, \tag{7}$$

where $C(b,c)$ is the normalizing coefficient.

In the following we focus our attention on symmetric states of Hamiltonian (4) only. In fact, as stressed by the authors of references [17),18)], this Hamiltonian couples symmetric states only with other symmetric states. Another symmetry of Hamiltonian (4) arises because the interaction moves pairs of particles only. Therefore H conserves the oddness or evenness of the population in each orbital. For M even the possibilities are *eee*, *eoo*, *ooe* and *oeo*, whereas for M odd they are *ooo*, *oee*, *eeo* and *eoe*.

3.2. – *Local Properties (Eigenvalue Statistics)*

In the classical studies of the transition from ordered to chaotic regime a useful tool is the examination of the phase space, in particular of the Poincaré sections [19)]. Such plots are *not* directly available in the case of quantal systems. One way of distinguishing order from chaos in such systems is to display the statistics of the level spacings in the energy spectrum. In fact, as discussed in references [20),21)], in the chaotic regime one expects a good agreement with GOE, whereas in the ordered regime results in agreement with Poisson statistic are expected.

The authors of references [17),18)] using the two statistics discussed in sect. 2 – $P(s)$ and Δ_3 – pointed out a sharp transition from Poisson to GOE predictions (see Figures 3,4,5,6).

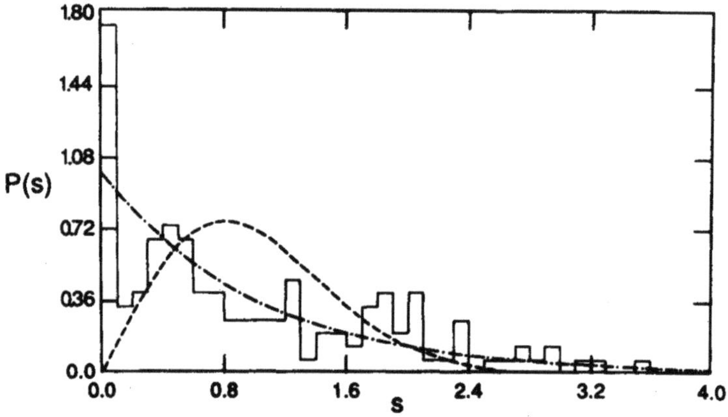

Figure 3. – Nearest–neighbour spacing distribution for $M = 102$, class *oeo* in the quasi-integrable region.

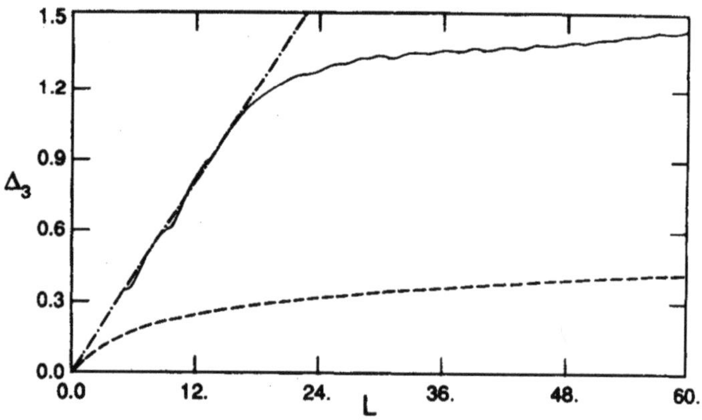

Figure 4. – Δ_3 statistic for the same states shown in Figure 3.

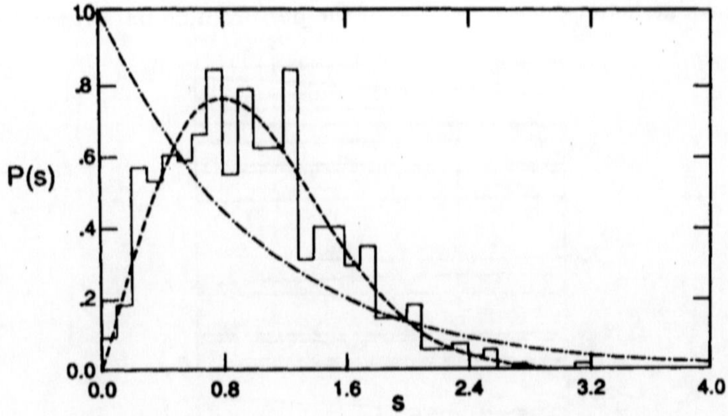

Figure 5. – Nearest–neighbour spacing distribution for $M = 102$, class *oeo* in the chaotic region.

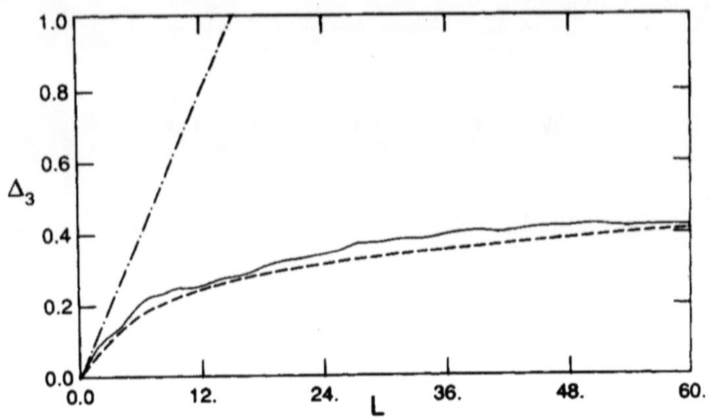

Figure 6. – Δ_3 statistic for the same states shown in Figure 5.

Another way of distinguishing a regular from a chaotic region is to study the eigenvalue behaviour as function of the perturbation parameter [22].

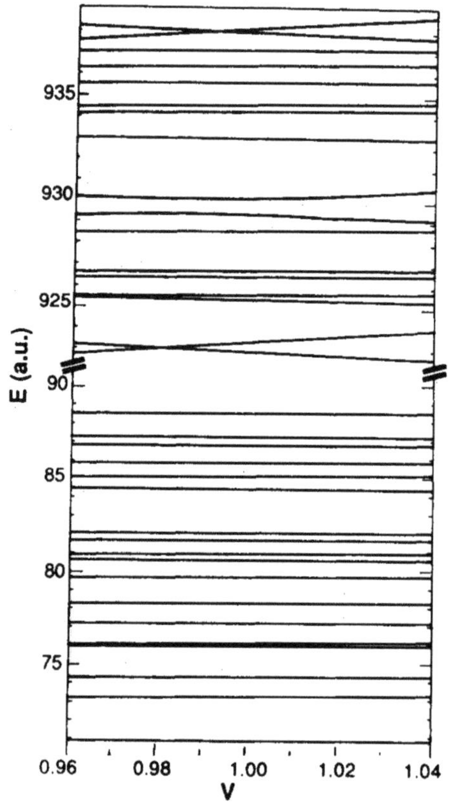

Figure 7. – Eigenvalue behaviour *vs* strength V of the two–body interaction in two regions of excitation spectrum; adapted from reference [24].

In Figure 7 two sets of levels, in regular and chaotic regions respectively, are shown, *vs* the strength of the interaction V [(*)] , whereas in Figure 8 the

(*) In order to remove the secular variation of the level density, each spectrum is mapped, by a numerical procedure described in reference [23], into one which has a constant level density.

second difference Δ, defined in the usual way [22]

$$\Delta \equiv \mid E(V+\Delta V) - 2E(V) + E(V-\Delta V) \mid; \qquad \Delta V = 0.01 \qquad (8)$$

is plotted *vs* E. Differences in the numerical values of Δ in the two regions of the excitation spectrum are very sharp.

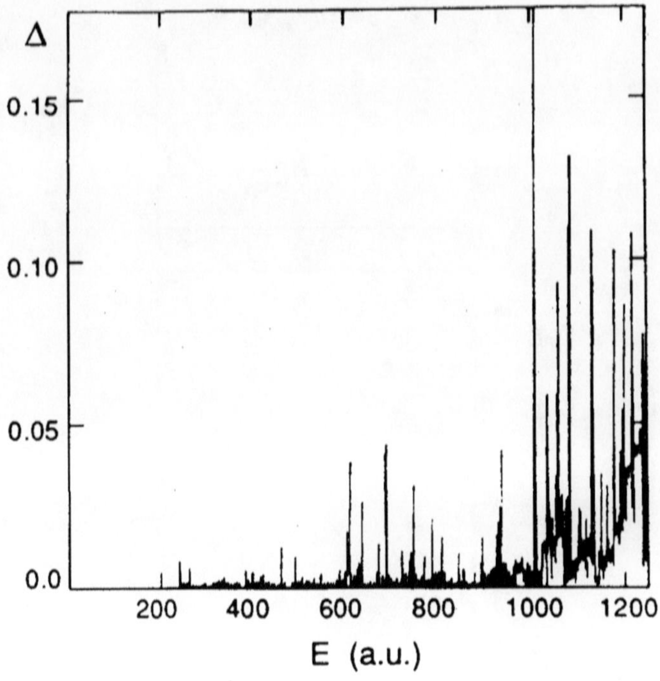

Figure 8. – The second difference Δ *vs* E; adapted from reference [24].

3.3 – *Wave Function Behaviours*

In order to point out the different behaviours of wave functions in regular and chaotic regions, two different measures have been used [25].

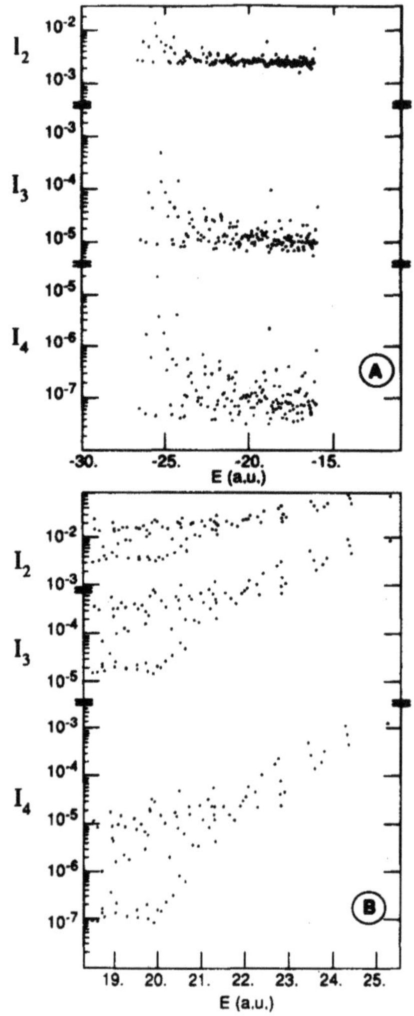

Figure 9. – Moments I_m *vs* energy E for $M = 102$, class *oeo*. The total number of states is 1326. In the figure only the moments corresponding to 150 states in the chaotic region (A) and 100 states in the quasi-integrable region (B) are shown; adapted from reference [25].

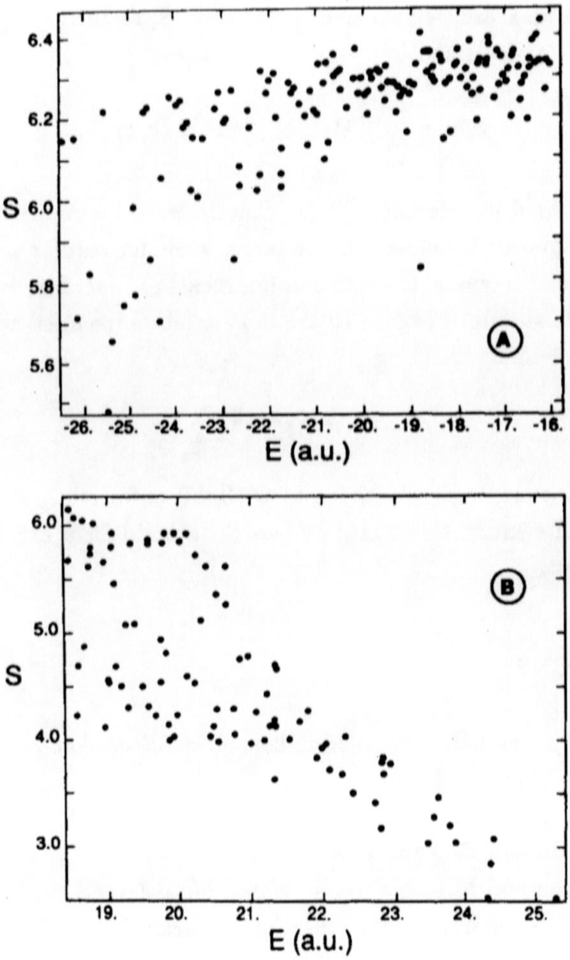

Figure 10. – Function S *vs* energy E for the same states shown in Figure 9; adapted from reference [25]. Figure 9 shows the low order intensity moments I_m, defined in the usual way [26]

$$I_m = \sum_{b,c} | \Psi(b,c) |^{2m} \tag{9}$$

in the two regions. As can be seen, the moments $m \geq 2$ diverge in the quasi–integrable region, whereas in the chaotic region show constant behaviour, in good agreement with the theory developed by Berry and co–workers [27),28)].

Another measure we have used is function S, recently proposed by some authors [26),29)] and defined by

$$S = -\sum_{b,c} \mid \Psi(b,c) \mid^2 \ell n \mid \Psi(b,c) \mid^2 . \tag{10}$$

As discussed in reference [26)] for chaotic wave functions one would expect the information measure S to be large, while for regular wave functions, localized in small regions, the information measure should be small. The numerical results shown in Figure 10 are in excellent agreement with the above predictions.

****

The author is deeply indebted to O. Bohigas for many suggestions and for his kind interest in this work, and to F. Iachello for a careful reading of the manuscript.

REFERENCES

1. O. Bohigas and H.A. Weidenmüller, *Ann. Rev. Nucl. Part. Science*, **38**, 421 (1988).
2. H.A. Weidenmüller, in *Heavy Ion Collisions*, Ed. by P. Bonche, M. Levy, P. Quentin and D. Vautherin (1986).
3. F.J. Dyson and M.L. Mehta, *J. Math. Phys.* (N.Y.) **4**, 701 (1963).
4. M.L. Mehta, in *Statistical Properties of Nuclei*, Ed. by J.B. Garg (1972).
5. N. Bohr, *Nature* **137**, 344 (1936).
6. F.J. Dyson, *J. Math. Phys.* **3**, 140 (1962).
7. E.P. Wigner, *SIAM Review* **9**, 1 (1967).
8. T. Ericson and T. Mayer–Kuckuk, *Ann. Rev. Nucl. Part. Sci.* **16**, 183 (1966).
9. H.I. Liou, G. Hacken, J. Rainwater and U.N. Singh, *Phys. Rev.* **C11**, 462 (1975).
10. G.E. Mitchell, in *Quantum Chaos and Statistical Nuclear Physics*, Ed. by T.H. Seligman and H. Nishioka (1986).
11. R.U. Haq, A. Pandey and O. Bohigas, *Phys. Rev. Lett.* **48**, 1086 (1982).

12. O. Bohigas, R.U. Haq and A. Pandey, in *Nuclear Data for Science and Technology*, Ed. by K.H. Böckhoff (1983).
13. M.L. Metha, *Random Matrices* (1967).
14. O. Bohigas, M.J. Giannoni and C. Schmit, *Phys. Rev. Lett.* **52**, 1 (1984).
15. O. Bohigas, M.J. Giannoni and C. Schmit, in *Chaotic Behaviour in Quantum Systems*, Ed. by G. Casati (1985).
16. H.J. Lipkin, N. Meshkov and A.J. Glick, Nucl. Phys. **62**, 188 (1965).
17. D.C. Meredith and S.E. Koonin, in *Quantum Chaos and Statistical Nuclear Physics*, Ed. by T.H. Seligman and H. Nishioka (1986).
18. D.C. Meredith, S.E. Koonin and M.R. Zirnbauer, Phys. Rev. **A37**, 3499 (1988).
19. M. Tabor, *Chaos and Integrability in Nonlinear Dynamics* (1989).
20. M.V. Berry, *Ann. of Phys.* **131**, 163 (1981).
21. M.V. Berry and M. Tabor, *Proc. R. Soc.* A**356**, 375 (1977).
22. I.C. Percival, *Advances in Chemical Physics*, **36**, 1 (1977).
23. V.R. Manfredi, *Lett. Nuovo Cimento*, **40**, 135 (1984).
24. L. Demattè, V.R. Manfredi and S. Montagnani, *Global and Local Properties of the Three Level Lipkin–Meshkov–Glick Model*, to be published.
25. M.T. Lopez–Arias and V.R. Manfredi, *Wave Function Behaviours of the Three-Level Lipkin, Meshkov, Glick Model*, Zeitschrift für Physik **A**, to be published.
26. J. Reichl, *Europhys. Lett.* **6**, 669 (1988).
27. M.V. Berry, *J. Phys.*, **A10**, 2083 (1977).
28. M.V. Berry, J.H. Hannay and A.M. Ozorio de Almeida, *Physica*, **D8**, 229 (1983).
29. F. Iachello and R.D. Levine, Europhys. Lett. **4**, 389 (1987).

PART II: TRANSITIONS FROM ORDER TO CHAOS IN PHYSICAL SYSTEMS.

ORDER OUT OF CHAOS IN ARITHMETIC

A. De Matteis

Laboratorio Matematica Applicata, ENEA, Bologna

Introduction

An interesting transition from deterministic chaos to order can be observed in number theory. This is a field characterized by the lack of continuity, where the knowledge of any non-trivial function (e.g., the Euler function or the greatest common divisor) inside a certain domain, does not allow any extrapolation, say, along its boundary. This unpredictable behaviour of arithmetical functions has such deep roots that it has recently been related by Chaitin |1| to the proof of incompleteness of arithmetic. His argumentation is based on the proof of existence of certain Diophantine equations depending on a parameter with the following property: when the parameter runs through all positive integers, we may find (score in this case H) or not (score T) an infinite number of solutions with a sequence of affirmative and negative answers (i.e. a sequence of H's and T's) indistinguishable from the outcomes of a fair coin. In the words of Chaitin, we are here faced with an arithmetic problem whose answer "jumps around in a completely unpredictable manner".

Due to this dominant character, since the very beginning of computer history |2| number theory has yielded the tools for introducing randomness in digital simulation. Our example, indeed, originates from stochastic simulation, namely, from the heart of simulation, i.e., random number generation; it has also a practical relevance in computational methods because it points out the appearance of strong correlations in sources of random (or, better, pseudo-random) numbers.

Power residues

If p is a prime number, then for any integer a Fermat's theorem states that

$$a^p \equiv a(\bmod p).$$

This implies that the sequence of the least non-negative remainders modulo p of the powers $a, a^2, a^3, \ldots$, is periodic and the least period g is $p-1$ or one of its divisors. Numbers a such that $g = p-1$ are called "primitive roots" of p ; the power residues modulo p of a primitive root run through all positive integers less than p , in an order which is generally different from the natural one. Finally, every odd prime has primitive roots |3|.

Assume now that a be a primitive root of p and consider in the system of the least residues modulo p the set of pairs (a^n, a^{n+d}) , $n = 1,2,\ldots$, i.e. the points in the plane whose coordinates are residues of powers with exponents at constant distance d . Fig.1 shows one thousand pairs of this type, normalised in the unit square, having chosen

$$p = 2^{31} - 1, \qquad a = 7^5, \quad d = (p-1)/2 - 1,$$

i.e., a large Mersenne prime as modulus, one of its primitive roots, and one unity less than half-period as distance between the exponents. The point distribution exhibits a chaotic character (really, we are in presence of a complex order, because the points lie on a very large number of parallel lines) which does not allow

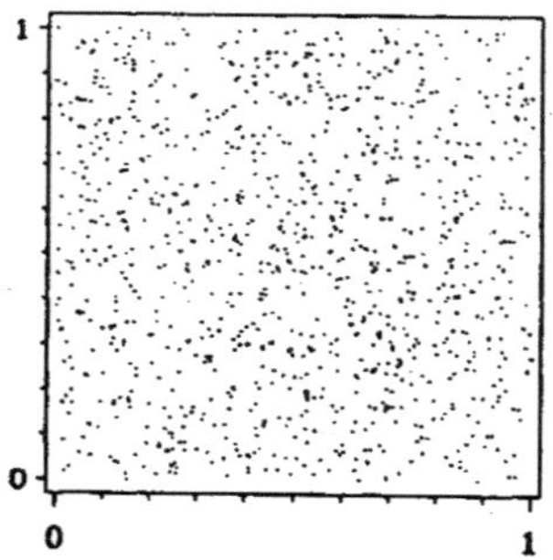

fig. 1 $d = (p-1)/2 - 1$

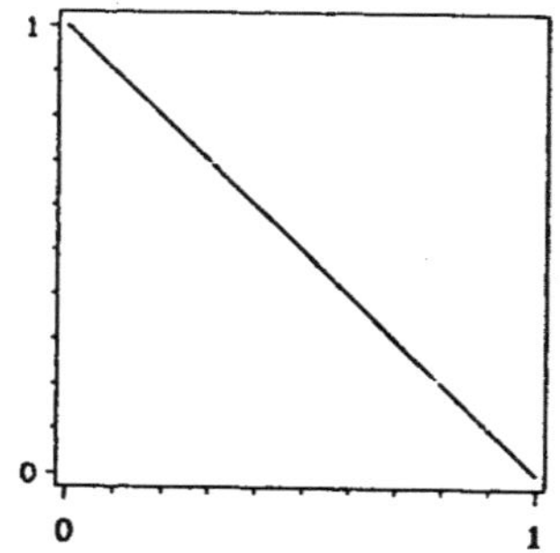

fig. 2 $d = (p-1)/2$

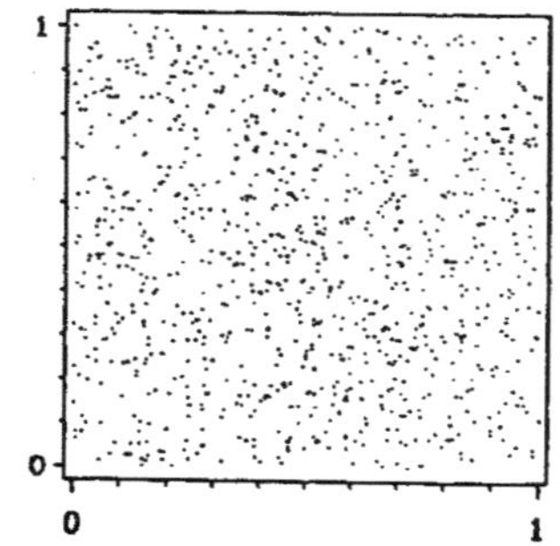

fig. 3 $d = (p-1)/2 + 1$

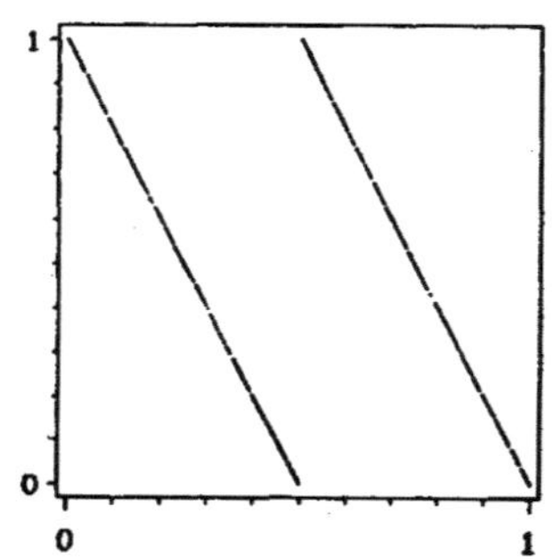

fig. 4 d = 311.731.497

the order that is going to appear for $d = (p-1)/2$ to be forecast: as shown in Fig.2, all points (not only the represented ones) will fall upon a diagonal of the square. They will return chaotic at the next unit increment of d (Fig. 3). This behaviour is not generated by the particular pair (p,a) considered, but is an invariant for every primitive root of a prime |4|; and occurs not only at the distance of half-period, but also for other values of d |5|, although with different patterns (see, e.g., Fig. 4).

The above phenomenon may also be presented as a dynamic process, by considering the trajectory described at discrete time instants by the equations

$$x_n \equiv a x_{n-1} (\bmod p)$$
$$y_n \equiv a y_{n-1} (\bmod p),$$

with the initial conditions

$$x_o = 1$$
$$y_o \equiv a^d (\bmod p).$$

This process is strongly sensitive to minimal variations of the starting value of the y-coordinate: when d changes, the trajectory of the point in the square may reduce to jumps along a line.

A Diophantine equation associated to the problem

The critical nature of the distance $d = (p-1)/2$, is a consequence of the relation (|4|, Lemma 7.1)

$$a^{(p-1)/2} \equiv (p-1) \ (\bmod p),$$

or, equivalently, of the fact that equation

$$x^{(m-1)/2} \equiv (m-1) \ (\bmod m),$$

is satisfied by any primitive root of the modulus m when m is prime. This last equation may be rewritten as

$$x^n \equiv 2n \ (\bmod 2n+1),$$

equivalent to the Diophantine equation containing the parameter n

$$x^n = (2n+1)y + 2n,$$

which has infinite solutions if and only if $m = 2n+1$ is prime. For each n we may thus decide in a finite number of steps (of increasing complexity) whether there are infinite solutions or not. With the conventions in |1| (write the sequence of answers as a real positive less than 1, with the digit 1 standing for H and 0 for T), the real number associated to the Diophantine equation considered is computable, and its first binary digits are

$$r = 0.11101101101001...$$

This means that the above sequence cannot be taken as independent tosses of a fair coin (prime number theorem). In other words, our Diophantine equation does not belong to the class of Chaitin.

Residues as pseudo-random numbers

Lehmer |2| suggested in 1949 the adoption of power residues as pseudo-random numbers in digital simulation of stochastic problems. They are generated by the recursive relation

$$x_n \equiv a x_{n-1} \pmod{p} \qquad n = 1,2,....$$

(subsequently normalized in the unit interval through division by p), starting from $x_0 = 1$. In the case of a prime modulus and of a primitive root a, the change of the starting value produces a cyclic permutation of the previous sequence. In more general cases |6|, a different initialization may produce a completely new sequence, with no common terms (a further example of extreme sensitivity to initial conditions).

The appearance of order in the dynamic process described above denounces a strong long-range correlation between terms far apart in the sequence: the manifestation of this correlation is sudden, without any premonitory sign, at certain distances d which must be determined in order to defend oneself. In the above examples we have met the critical half-period distance, which inhibits the equipartition of the sequence between two concurrent processors of a parallel computer (the phenomenon was first observed during the attempts to parallelize Monte Carlo computations and the numerical case considered is the IBM-3090 generator |7|).

Long-range correlations are inherent in the generation algorithm, implying that they cannot be counteracted by judicious choices of the parameters appearing in the recursive formula, but only avoided by drastic reduction of the sequence length usable. They originate at long distance, but may propagate - although attenuated - towards shorter distances |8|. Finally, they also appear between disjoint sequences, i.e., one has to face not only autocorrelations but also mutual correlations when using more than one generator |9|.

A final remark

An example of order out of disorder has been shown in the theory of linear congruences. We have also commented upon the unpredictable character of arithmetical functions and the unpredictability of the type of solution in certain Diophantine equations. There is a further aspect in the practical implementation of arithmetical operations on digital computers which we would briefly comment since it may cast a shadow on the affidability of a tool which is fundamentally arithmetic - the digital computer - particularly in the simulation of nonlinear dynamic problems and in the investigation of the parameters giving rise to the chaos.

The algebraic structure of "floating point arithmetic", i.e. of the numbers representable in a real computer (with a fixed number of significant digits) under floating point operations, appears almost evanescent: it may be sufficient to note that addition is not associative (think of the sum $b^n + 1 - b^n$ in a floating point machine with base of representation b and n digits for the mantissa) to go down in the hierarchy - as far as addition is concerned - below the semigroup. Equipped with these arms, the greatest caution seems necessary in the numerical simulation of systems approaching order-chaos transitions, because we cannot exclude that our "arithmetic" may inject its own chaos into the studied system (one such example has been constructed in |10|).

It is tempting to compare the undesired disorder which may be produced by this weak structure with the undesired order appearing in the robust structure (no less than a field) of the remainders modulo p under addition and multiplication.

Acknowledgements - The author is indebted to Professor Mario Rasetti who brought to his attention the work of Chaitin. He also wishes to thank his colleague Daniele Visparelli who enlived this talk with a computer animated presentation of the dynamic process.

References

1. G.J. Chaitin "Incompleteness theorems for random reals" Adv. in Appl. Math. 8 (1987) 119-146.

2. D.H. Lehmer "Mathematical methods in large-scale computing units" in Symposium on Large Scale Calculating Machinery, Harvard University, 1949, 141-146.

3. I.M. Vinogradov, Elements of Number Theory, Dover Publications, New York, 1954.

4. A. De Matteis, S. Pagnutti "Parallelization of random number generators and long-range correlations" Numer.Math. 53 (1988) 595-608.

5. J. Eichenauer-Herrmann, H. Grothe "A remark on long-range correlations in multiplicative congruential pseudo-random number generators" Numer. Math. 56 (1989) 609-611.

6. G. Cenacchi, A. De Matteis "Pseudo-random numbers for comparative Monte Carlo calculations" Numer.Math. 16 (1970) 11-15.

7. Engineering and Scientific Subroutine Library Guide and Reference, SC23-0184-2, p. 763 , IBM Corporation (1987).

8. A. De Matteis, S. Pagnutti "Long-range correlations in linear and non-linear random number generators" Parallel Computing (1990,in press)

9. A. De Matteis, S. Pagnutti "A class of parallel random number generators" Parallel Comput. 13 (1990) 193-198.

10. A. De Matteis "Improvements based on computed errors" Comm. ACM 10 (1967) 683.

CHAOTIC DYNAMICS OF LASER OSCILLATORS

F.T. Arecchi
Dept. of Physics, University of Florence and
Istituto Nazionale di Ottica, Firenze

ABSTRACT

This is a review of how deterministic chaos enters quantum optics, introducing new features unexpected in the first twenty years of the laser era. After a general introduction to chaos in lasers, attention is focused to CO_2 lasers with feedback, which display the so called Shil'nikov chaos.

1. Introduction

Quantum optics from its beginning in 1960 with the first laser was considered as the physics of coherent and intrinsically stable radiation sources. Lamb's semiclassical theory[1] showed the role of the EM field in the cavity in ordering the phases of the induced atomic dipoles, thus giving rise to a macroscopic polarization and making possible a description in terms of very few collective variables. In the case of a single-mode laser and a homogeneous-gain line this meant just five coupled degrees of freedom, namely, a complex field amplitude E, a complex polarization P, and a population inversion ΔN. A corresponding quantum theory, even for the simplest model laser (the so called Dicke model, that is, a discrete collection of modes interacting with a finite number of two-level atoms) does not lead to a closed set of equations. However the interaction with other degrees of freedom acting as a thermal bath (atomic collisions, thermal radiation) provides truncation of high-order terms in the atom-field interation. The problem may be reduced to five coupled equations (the so-called Maxwell-Bloch equations) but now they are affected by noise sources to account for the coupling with the thermal bath. Being stochastic, or Langevin equations, the corresponding solution in closed form refers to a suitable weight function or phase-space density. Anyway the average motion matches the semiclassical one, and fluctuations play a negligible role if one excludes the bifurcation points where there are changes of stability in the stationary branches. Leaving out the peculiar statistical phenomena which characterize the threshold points and which suggest a formal analogy with thermodynamic phase transitions,[2] the main point of interest is that a single-mode laser provides a highly stable or coherent radiation field. From the point of view of the associated information, the standard interferometric or spectroscopic measurements of classical optics, relying on average field values or on their first-order correlation functions, are insufficient. In order to characterize the statistical features of Quantum Optics it was necessary to make extensive use of photon statistics[3,4].

Coherence is equivalent to having a stable fixed point attractor and this does not depend on details of the nonlinear coupling, but on the number of relevant degrees of freedom. Since such a number depends on the time scales on which the output field is observed, coherence becomes a question of time scales. This is the reason why for some lasers coherence is a robust quality, persistent even in the presence of strong perturbations, whereas in other cases coherence is easily destroyed by the manipulations common in the laboratory use of lasers, such as modulation, feedback or injection from another laser.

Sect. 2 is a general presentation of low-dimensional chaos in lasers, including the description of the relevant measurements upon which any assessment on chaos has to rely. Sect. 3 describes features of a new type of chaos, the Shil'nikov chaos.

2. Deterministic chaos

Until recently the current point of view was that a few-body dynamics was fully predictable, and that only addition of noise sources, due to coupling with a thermal reservoir, could provide statistical fluctuations. Lack of long-time predictability, or turbulence, was considered as resulting from the interaction of a large number of degrees of freedom, as in a fluid above the critical Reynolds number (Landau-Hopf model of turbulence).

On the contrary, it is now known that even in systems with few degrees of freedom nonlinearities may give rise to expanding directions in phase space and this, together with the lack of precision in assigning initial conditions, is sufficient to induce a loss of predictability over long times.

This level of dynamical description was introduced by Poincaré, in dealing with the three-body problem in celestial mechanics. Already a three-body dynamic system is very different from the two-body problem since, in general, it may include an asymptotic instability. This means a divergence, exponential in time, of two phase space trajectories stemming from nearby initial points. The uniqueness theorem for solutions of differential systems always holds once one has assigned the coordinates of the initial point. However a fundamental difficulty arises. Only rational numbers can be assigned by a finite number of digits. A "precise" assignment of a real number requires an infinite acquisition time and an infinite memory capacity to store it, and neither of these two infinities is available to the physicist. Hence any initial condition implies a truncation. A whole range of initial conditions, even if small, is usually given and from within it trajectories may arise whose difference becomes sizeable after a given time, if there is an exponential divergence. As consequence predictions are in general limited in time and motions are complex, starting already from the three-body case. In fact, we know nowadays from very elementary topological considerations that a three-dimensional phase space corresponding to three coupled degrees of freedom is already sufficient to yield a positive Lyapunov exponent, and accordingly an expanding phase-space direction. This complexity is not due to coupling with a noise source as a thermal reservoir, but to sensitive dependence on initial conditions. It is called deterministic chaos.

2.1. Dynamical aspects of chaos

A dissipative system (i.e. with damping terms) does not conserve the phase-space volume. If we start with initial conditions confined in a hypersphere of radius ε, that is, with an initial phase volume

$$V_0 = \varepsilon^N$$

as time goes on, the sphere transforms into an ellipsoid with each axis modified by a time dependent factor. Its volume is

$$V_t = \varepsilon^N \exp(\Sigma_i \lambda_i t)$$

(λ_i: Lyapunov exponents). Since the volume has to contract, $V_t < V_0$, then

$$\Sigma_i \lambda_i < 0$$

We denote the sequence of λ exponents, starting from the smallest up to the highest as the Lyapunov spectrum. Let us consider for simplicity just the signs of nonzero λ_i, keeping the zero for $\lambda_i = 0$. We then describe a sequence of, negative, zero and positive λ_i as, e.g., (- - 0 +). For N = 1, we have (-) and a segment $V_0 = \varepsilon^1$ of initial conditions shrinks to a single point for $t \to \infty$, that is, the attractor is a fixed point. For N = 2 the system starts from $V_0 = \varepsilon^2$ and goes either to a fixed point (- -), or to a limit cycle (-0). Chaotic motion (-+) with $\lambda_+ < |\lambda_-|$ is forbidden in two dimensions by the Poincaré-Benedixon theorem. For N = 3, besides fixed point (- - -), and limit cycle (- - 0), we can have motion on a torus with two incommensurate frequencies (- 0 0), but we can also have (- 0 +), that is, a positive λ which gives an expanding direction along which we rapidly get uncertainty.

An example of chaotic motion is offered by the Lorenz model of hydrodynamic instabilities which corresponds to the following equations where the parameter values have been chosen in order to yield one positive Lyapunov exponent:

$$
\begin{aligned}
\dot{x} &= 10x + 10y, \\
\dot{y} &= y + 28x - xz, \\
\dot{z} &= -(8/3)z + xy.
\end{aligned}
\tag{1}
$$

The above considerations suggest the system will exhibit low-dimensional chaos, with the simplest phase-space topology allowing for the appearence of a positive Lyapunov exponent.

Focusing on these situations in quantum optics permits close comparison between experiments and theory. By purpose, I do not tackle the vast class of inhomogeneously broadened lasers, where it is extremely difficult to drive close correspondences between experiments and theory because of the large number of coupled degrees of freedom.

If we couple Maxwell equations with Schrödinger equations for N atoms confined in a cavity, and expand the field in cavity modes, keeping only the first mode E which goes unstable, this is coupled with the collective variables P and Δ describing the atomic polarization and population inversion as follows,

$$
\begin{aligned}
\dot{E} &= -kE + gP, \\
\dot{P} &= -\gamma_{\perp} P + gE\Delta, \\
\dot{\Delta} &= -\gamma_{\parallel}(\Delta - \Delta_0) - 4gPE.
\end{aligned}
\tag{2}
$$

For simplicity, we consider the cavity frequency at resonance with the atomic resonance, so that we can take E and P as real variables and we have three coupled equations. Here, k, $\gamma_{\perp}$, $\gamma_{\parallel}$ are the loss rates for field, polarization and population, respectively, g is a coupling constant and Δ_0 is the population inversion which would be established by the pump mechanism in the atomic medium, in the absence of coupling. While the first equation comes from Maxwell equations, the two others imply the reduction of each atom to a two-level atom resonantly coupled with the field, that is, a description of each atom in a isospin space of spin 1/2. The last two equations are like Bloch equations which describe the spin procession in presence of a magnetic field.

The similarity of Maxwell-Bloch equations with Lorenz equations would suggest the easy appearence of chaotic instabilities in single-mode, homogeneous-line lasers. However, time-scale considerations rule out the full dynamics of (2) for most of the available lasers. Equations (1) have damping rates which lie within one order of magnitude of each other. On the contrary, in most lasers the three damping rates are wildly different from one another.

The following classification has been introduced

Class A (e.g., He-Ne, Ar, Kr, dye): $\gamma_{\perp} \simeq \gamma_{\parallel} \gg$ k. The last equations of (2) can be solved at equilibrium (adiabatic elimination procedure), thus the laser dynamics is described by one single nonlinear field equation. N = 1 means fixed point attractor, hence coherent emission.

Class B (e.g., ruby, Nd, CO): $\gamma_\perp \gg k \gtrsim \gamma_\parallel$. Only polarization is adiabatically eliminated (middle equation of (2)) and the dynamics is ruled by two rate equations for field and population. N = 2 allows also for periodic oscillations.

Class C (e.g., FIR lasers) $\gamma_\parallel \simeq \gamma_\perp \simeq k$. The complete set of (2) has to be used, hence Lorenz like chaos is feasible.

We have carried a series of experiments on the birth of deterministic chaos in CO_2 lasers (class B). In order to increase, by at least 1, the number of degrees of freedom, we have tested the following configurations:

i) Introduction of a time dependent parameter to make the system non autonomous. Precisely, an electro-optical modulator modulates the cavity losses at a frequency near the proper oscillation frequency Ω provided by a linear stability analysis, which for a CO_2 laser happens to lie in the 50-100 KHz range, providing easy and accurate sets of measurements.

ii) Injection of a signal from an external laser detuned with respect to main one, choosing tne frequency difference near the above mentioned Ω . With respect to the external reference the laser field has two quadrature components which represent two dynamical variables. Hence we reach N = 3 and observe chaos.

iii) Use a bidirectional ring rather than a Fabry-Perot cavity. In the latter case the boundary conditions constrain the forward and backward waves, by phase relations on the mirror, to act as a single standing wave. In the former case forward and backward waves have just to fill the total ring length with an integer number of wavelengths but there are no mutual phase constrains, hence they act as two separate variables. Furthermore, when the field frequency is detuned with respect to the center of the gain line, a complex population grating arises from interference of the two counter-going waves, and as a result the dynamics becomes rather complex, requiring $N > 3$ dimensions.

iv) Add an overall feedback, besides that provided by the cavity mirrors, by modulating the losses with a signal provided by the output intensity. If the feedback has a time constant comparable with the population decay time, it provides a third equation sufficient ot yield chaos.

Notice that while methods (i), (ii) and (iv) require an external device, (iii) provides intrinsic chaos. In any case, since feedback, injection or modulation are currently used in laser applications, the evidence of chaotic regions puts a caution on the optimistic trust in the laser coherence.

2.2. Information aspects of chaos

Here we discuss what we measure to characterize chaos. We plot two of the three (or more) variables on a plane phase-space projection. This way, we build projections of phase space trajectories on an x-y oscilloscope. Simultaneously we can measure the power spectrum. In first experiment on a chaotic laser a sequence of subharmonic bifurcations was shown, which eventually leads to an intricated trajectory (strange attractor) and to a continuous power spectrum. But how can we discriminate between deterministic chaos and noise? After all, noise also would give a continuous spectrum, and the phase space point would fill ergodically part of the plane, thus densely covering a two-dimensional set.

In order to discriminate deterministic chaos from order as well as from random noise, we introduce two invariants of the motion, one static the other dynamic.

We partition the phase space into small boxes of linear size $\mathcal{E}$ and give ith box a probability $p_i = M_i/M$ equal to fractional number of times it has been visited by the trajectory. This way, we build a Shannon information $I(\mathcal{E})$, and with it an "information dimension" $D_1(\mathcal{E})$ which is, in general, a fractional number, or a "fractal".

$$I(\varepsilon) = -\sum_i p_i \log p_i \,, \tag{3}$$

$$D_1(\varepsilon) = -\lim_{\varepsilon \to 0} \frac{I(\varepsilon)}{\log \varepsilon} \,. \tag{4}$$

To understand the meaning of a fractal, look up an operational definition of dimension. Let us compare three sets: (i) a segment of unit length; (ii) the Cantor set, built by taking out the middle one-third of the unit segment and repeating the operation on each fragment; (iii) the Koch curve, built by replacing the middle third with the other two sides of an equiteral triangle and repeating the operation "ad infinitum". At each stage of the partition, we cover each set with beads of suitable size not to lose in resolution (e.g., diameter 1/3 at the first partition) and count the number N for each set (at the first partition, we need 2 for the Cantor set, 3 for the segment, 4 for the Koch curve).

We define the fractal dimension as the ratio

$$D_0(\varepsilon) = \frac{\log N(\varepsilon)}{\log 1/\varepsilon} \,. \tag{5}$$

This definition is independent of the partition. Indeed, for the Cantor set and the Koch curve we have $N = 2$, $\mathcal{E} = 1/3$ and $N = 4$, $\mathcal{E} = 1/3$, respectively, at the first partition, yielding

$$D_0(\text{Cantor}) = \frac{\log 2}{\log 3} \simeq 0.63..., \quad \text{and}$$

$$D_0(\text{koch}) = \frac{\log 4}{\log 3} \simeq 1.2618... \ .$$

At the second partition the number of necessary beads goes as N^2 and the diameter of each as ε^2, hence D_0 remains invariant.
Going back to the information dimension $D_1(\varepsilon)$ we see that we have replaced log N with $I(\varepsilon)$ which is an average (for p_i all equal, we recover $I(\varepsilon) = \log N$). Hence D_1 generalizes D_0 whenever the density of points is not uniform along the trajectory.
As D_0 was independent of the partition stage, similarly D_1 is an invariant, but static (time does not enter). It can be shown that $D_0 \geq D_1$, however for non pathological sets the difference is irrelevant. Let us refer for simplicity to an N = 3-dimensional phase space. If D = 0 (fixed point) or 1 (limit cycle) or 2 (torus) we have an ordered, or coherent, motion. In the order limit of random noise, fluctuations fill ergodically an N dimensional region of the space, hence D = 3. Deterministic chaos has to be in between, that is

$$2 < D < 3.$$

Hence, a fractal dimension is an indicator of chaos.
These features related to the topology of the attractor have a temporal counterpart in another invariant, which measures how information is dissipated in a motion to maintain knowledge of the system. To build this dynamic invariant, we partition both space and time in boxes of sizes ε and τ that we name $i_1, i_2, \ldots i_d$ at each of the discrete times τ, $2\tau, \ldots$ $d\tau$, and introduce the joint probability over the d time intervals,

$$P_{i_1 i_2 \ldots i_d} \equiv \left\{ x(t = \tau) \subset i_1 \ ; \ \ldots \ ; \ (t = d\tau) \subset i_d \right\}$$

Correspondingly, we define a joint information

$$I_d(\varepsilon) = - \sum_{\{i_1 \ldots i_d\}} p_{i_1 \ldots i_d} \log p_{i_1 \ldots i_d} \ . \tag{6}$$

Then, by a limit operation, define the Kolmogorov entropy as the rate of information loss per unit time

$$K \equiv \lim_{\tau \to 0} \lim_{\varepsilon \to 0} \lim_{d \to \infty} \frac{1}{d\tau} \sum_{n=1}^{d} (I_{n+1} - I_n) = \lim \frac{1}{d\tau} I_d \ . \tag{7}$$

Now we have two indicators to gauge the difference among order, random noise (Brownian motion) and deterministic chaos. Referring to K, it is easily seen that

$K = 0$ for order (no information loss);

$K = \infty$ for random noise (total information loss);

$0 < K < \infty$ for deterministic chaos.

The box counting method described above is impractical. It may require 10^6 points for a convergent numerical result. On the contrary, the following method introduced by Grassberger and Procaccia is applicable to only 10^3-10^4 independent data points. We generalize Shannon information defining the order-f information as

$$I_f(\varepsilon) = \frac{1}{1-f} \ln \sum_i p_i^f . \tag{8}$$

For $f \to 1$ we recover the usual definition. Associated with I_f, there is an order-f dimension of the attractor

$$D_f = \lim_{\varepsilon \to 0} \frac{I_f(\varepsilon)}{\ln 1/\varepsilon} . \tag{9}$$

For $f = 0$ and 1 we recover D_0 and D_1. Consider $f = 2$. The sum $\sum p_i^2$ is just the probability that a pair of random points on the attractor fall into the same box, that is, that two arbitrary points will have a distance less than ε. Calling this probability $C(\varepsilon)$, we expect thus

$$C(\varepsilon) = \lim_{N \to \infty} \frac{1}{N^2} \sum_{ij} \theta(\varepsilon - |x_i - x_j|) \simeq \varepsilon^{D_2} \tag{10}$$

where $C(\varepsilon)$ is measured as the number of pairs (i,j) with a distance $|x_1 - x_j| < \varepsilon$. In (10), θ is the Heaviside step function.
Experimentally, we do not measure at each time the vector x(t) of phase space, but just one component $x_i(t)$ (for instance, just the light out of a laser).
However, in a nonlinear system, any component $x_k(t)$ will influence x_i at a later time (no normal mode transformation!). Hence, we can build an

m-dimensional phase space $\xi(t)$ by just measuring one single component x_i at successive times and considering the m-fold as a single point in m space

$$\xi(t) \equiv \left[x_i(t), x_i(t+\tau), \ldots x_i(t + (m - 1)\tau)\right] . \tag{11}$$

As we evaluate the slope log C vs log ε from our data, we can stop from increasing m when the slope shows saturation. The saturated slope is D_2.

3. Experimental characterization of Shil'nikov chaos by statistics of return times

The dynamic behavior of a single-mode CO_2 laser with feedback is characterized by global features in the phase space, related to the presence of three coexisting unstable fixed points. As a control parameter is monotonically increased, one can observe transitions from a Hopf bifurcation to a local chaos and eventually to regular spiking and Shil'nikov chaos.

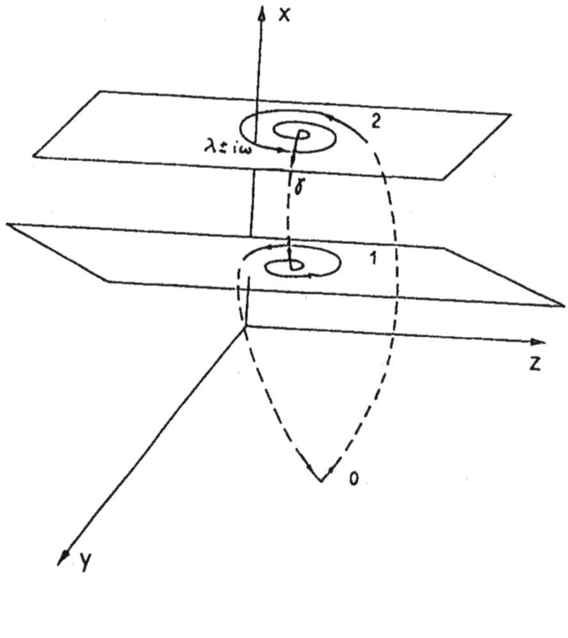

Fig. 1. Schematic view of a trajectory in the phase space when the dynamics are affected by all three unstable stationary points.

Furthermore, one can find evidence of competition among these different kinds of instability.[5] The phase-space trajectories are affected differently by each of the three unstable points, and by adjustment of the control parameters they can be characterized by the dominant role of only one, or a pair of them.

A linear stability analysis shows the local features at each fixed point. Precisely, point 0 (at zero intensity) is a saddle node with two stable directions and one unstable; point 1 has a plane unstable manifold with a focus and a stable third direction; point 2 has a stable manifold with a focus and an unstable third direction. Shil'nikov chaos is related to the saddle focus character of point 2. Around a saddle focus the motion consists of a contracting spiral $\exp(-\lambda t)\cos(\omega t)$ on the stable manifold and of an exponential expansion $\exp(\gamma t)$ along the unstable manifold. The presence of the other two unstable points ensures that the diverging flow is reinjected into the neighborhood of the saddle focus. Shil'nikov showed that for $|\lambda| < \gamma$ there exists a countable set of unstable trajectories close to the homoclinic one.[6] This structure of the flow is one of the simplest capable of generating chaotic behavior in many autonomous systems, such as the Lorenz equations[7] and the Belousov-Zhabotinski reaction.[8]

The temporal behavior of laser output intensity in this regime is characterized by pulses almost equal in shape but with chaotic recurrence times.[9] The regularity in shape means that the points at any Poincaré section are so closely packed that impossibly precise measurements of their position would be required if the relevant features of the motion were to be found. Instead, there is a large spread in the return times to a Poincaré section close to the unstable point. For this reason, the statistics of the return times appears to be the most appropriate characterization of Shil'nikov chaos.

Our experimental setup consists of a single mode CO_2 laser with an intracavity electro-optic modulator. A signal proportional to the laser output intensity is sent back to the electro-optic modulator[10]. Single mode CO_2 lasers have a dynamic behavior described by two coupled differential equations, one for the field amplitude and the other for the population inversion, the fast polarization being adiabatically eliminated from the complete set of Maxwell-Bloch equations. Thus, the presence of feedback introduces a third degree of freedom. When the feedback loop is so fast that it provides a practically instantly adapted loss coefficient, it does not modify the phase-space topology. On other hand, if the time scale of the feedback loop is of the same order as of the other two relevant variables, the system becomes three dimensional. With suitable normalizations such a system is described by three first-order differential equations for the laser intensity $x(t)$, the population inversion $y(t)$, and the modulation voltage $z(t)$ as follows:

$$
\begin{aligned}
x &= -K_0 x \left[1 + \alpha \sin^2(z) - y\right], \\
y &= -\gamma_{\parallel}(y + xy - A), \qquad\qquad (12) \\
z &= -\beta\,(z - B + \tau x),
\end{aligned}
$$

For a fixed pump A (fixed discharge current in the laser tube), our system has two control parameters: the bias voltage B applied to the electro-optic modulator and the gain r in the feedback loop.
From an experimental point of view we are able to visualize (x - z) phase-space projections, obtained by feeding onto a scope the photodetector signal proportional to the laser output intensity x(t) and the feedback voltage z(t). These phase-space projections consist of closed orbits visiting successively the neighborhoods of the three unstable stationary points 0,1, and 2.

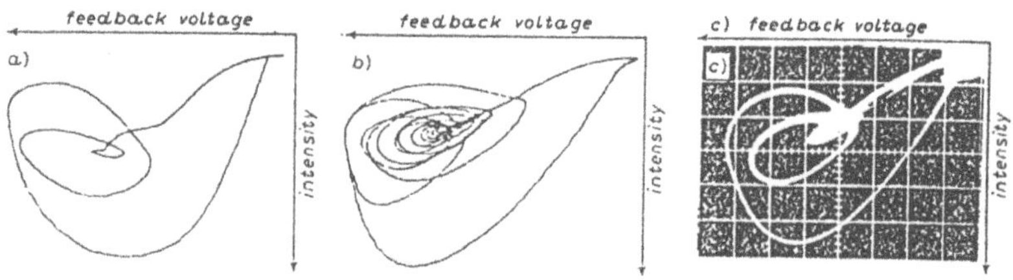

Fig. 2. Phase space projections x-z (laser intensity-feedback voltage). a) and b) are single orbits obtained by a digitizer, referring to the same parameters of Fig. 3a) and b), respectively. c) is the superposition of 30000 orbits of type a)

The local chaos around point 1, established at the end of a subharmonic sequence, has been characterized by standard methods as power spectra and correlation dimension measurements.[10]
However, the existence of a global behavior characterized by pulses with regular shapes but chaotic in their time of occurrence makes it significant to study the dynamics through measurements of return times to a Poincaré section. The measurements have been done by using a threshold circuit. An appropriate Poincaré section x = constant can be selected by adjusting the threshold level. This method permits us to distinguish among the different dynamical regimes.Adjusting the control parameters in order to have a dominance of the saddle focus 2, we obtain a motion consisting of a quasi-homoclinic orbit asymptotic to it (Fig. 3). In this regime, the laser output is characterized by pulses with regular shapes but chaotic in their recurrence. Based on such a consideration the iteration map of return times (τ_{i+1} versus τ_i) displays an extremely regular structure that we show below to be in close agreement with that arising from Shil'nikov theory of homoclinic chaos.
From a theoretical point of view, a homoclinic orbit asymptotic to a saddle focus can be modeled in terms of the following one-dimensional iteration map[11]:

$$\zeta_{n+1} = \zeta_n^{\lambda/\gamma} \cos[\omega/\gamma \ln(\zeta_n)] + \epsilon, \tag{13}$$

where γ and $-\lambda \pm i\omega$ are the eigenvalues of the linearized flow at the saddle focus, ζ is the coordinate along the unstable manifold, and ε is the deviation along ζ from the homoclinic orbit at the Poincaré section in the neighborhood of the saddle point (ε= 0 corresponds to the homoclinic condition).

If we build a small cubic box of unit side centered at the saddle focus and oriented along the eigenvectors ξ, η, and ζ , any tiny difference in the entrance coordinate along the expanding axis ζ will strongly influence the residence time inside the box and hence the spacing from the next reinjection.

Observing that most of the time is spent in the box around the saddle point, we relate the return time to the coordinate ζ of the unstable manifold by $\zeta = \zeta_0 \exp(\gamma\tau)$, thus obtaining an iteration map for the return times

$$\begin{aligned}\tau_{n+1} &= -\ln[\exp(-\lambda/\gamma\tau_n)\cos(\omega/\gamma\tau_n) + \epsilon] \\ &= -\ln[\varphi(\tau) + \epsilon],\end{aligned} \tag{14}$$

Comparison of Eqs.(13) and (14) shows the enhanced sensivity to fluctuations of the map with respect to the map.

Indeed, suppose that the offset ζ from homoclinicity is affected by a small amount of noise. The sensitivities of the two maps to such a noise are given, respectively, by $\partial\zeta/\partial\epsilon$= 1 and

$$\partial\tau/\partial\epsilon = [\varphi(\tau) + \epsilon]^{-1}. \tag{15}$$

This sensitivity factor acts as a lever arm whenever $\varphi(\tau) + \varepsilon$ becomes very small. Note the following: (1) This is not deterministic chaos; in fact, large fluctuations can be expected even for a regular dynamics, implying a fixed point τ^* . (2) It is not associated with the homoclinicity condition $\varepsilon = 0$; in fact, for finite ε there may be a τ^* such that $\varphi(\tau^*) + \varepsilon = 0$.

In figures 4 and 5 we show numerical and experimental iteration maps respectively. Fig. 5 also shows the spread related to the enhanced sensitivity to fluctuations.

To summarize, a CO_2 laser with feedback shows different dynamic regimes depending on the dominant role of one or two of three coexisting unstable stationary points. In particular, in the regime of Shil'nikov chaos the iteration maps of return times display a statistical spread owing to a transient fluctuation enhancement phenomenon peculiar to macroscopic systems, which is absent in low-dimensional chaotic dynamics.

In fact, the model description $\dot{x} = F(x)$ of a large system in terms of a low-dimensional dynamic variable x is just an ensemble-averaged description, and residual fluctuations on position x must be considered at some initial time, even though the successive evolution is accounted for by a deterministic law. In our case such a fluctuation is a stochastic spread $\delta\xi$ on the offset ξ of the position ζ.

The same amount of $\delta\xi$ in Eqs.(13) and (14) leaves the ζ maps unaltered, while it strongly affects the τ maps, making them appear like the experimental data.

We have thus shown a fundamental difference between a small system, ruled by a few equations, and a large system, in which the corresponding low-dimensional dynamics is a contracted description in terms of macroscopic variables, which are ensemble averages over some initial spread.

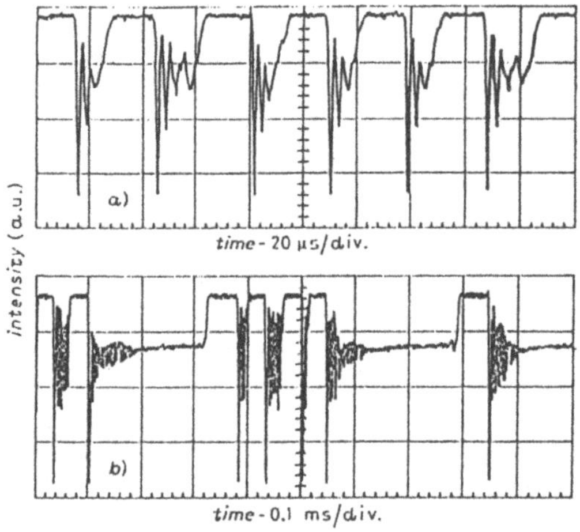

Fig. 3. Time plots of the intensity in the regime of Shil'nikov chaos. (a), (b) Refer to the same B value (B = 0.427) but two different gains of the feedback loop. (b) Shows long transients corresponding to a large number of small spirals around the saddle focus.

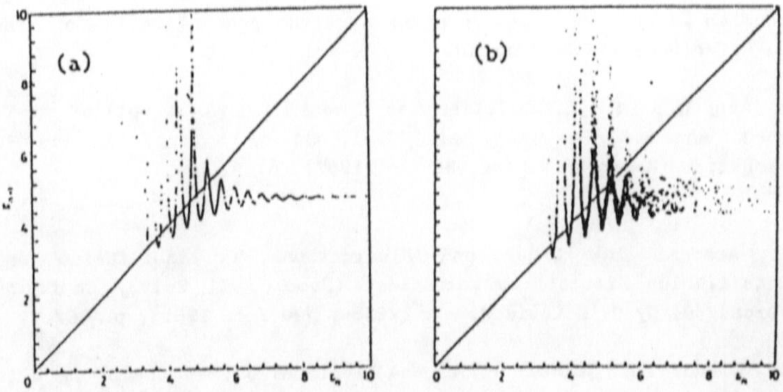

Fig. 4. Numerical iteration maps for Shil'nikov chaos. Paramater values: $\omega/\gamma = 13.0, \lambda/\gamma = 0.986, \mathcal{E} = 0.01$. (a) and (b), τ maps without and with noise $\delta\mathcal{E} = 10^{-2}$, respectively.

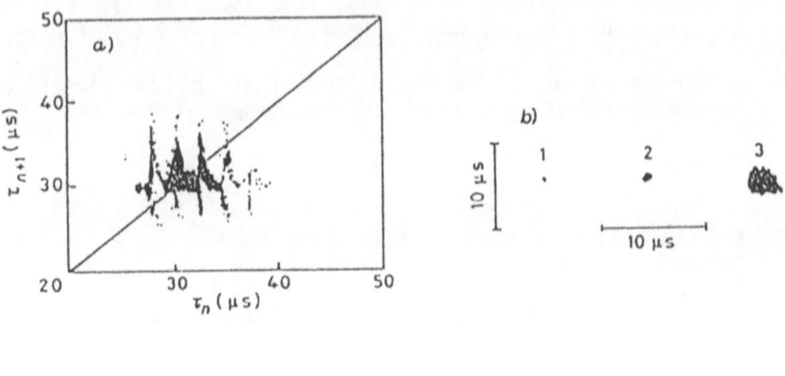

Fig. 5. Experimental iteration maps of the return times. (a) $\tau = 0.487$ and B = 0.350. (b) Maps corresponding to regular periodic situations, namely, 1, an electronic oscillator; 2, the laser in a regular periodic regime; 3, the laser just at the onset of the instability but still with a regular period.

REFERENCES

1. Rather than giving specific references to well known fundamental papers on laser physics, I refer to a recent general review of mine plus References listed therein.

F.T. Arecchi, in "Instabilities and Chaos in Quantum Optics", F.T. Arecchi and R.G. Harrison, eds., Vol. 34 of Springer Series in Synergetics (Springer-Verlag, Berlin, 1987), p. 9.

2. F.T. Arecchi in "Order and Fluctuations in Equilibrium and Nonequilibrium Statistical Mechanics" (Proc. XVII Solvay Conf. on Physics) ed. by G. Nicolis et al. (Wiley, New York 1981), p. 107.

3. R. J. Glauber in "Quantum Optics and Electronics", ed. by D. De Witt et al. (Gordon and Breach, New York 1965).

4. F.T. Arecchi in "Quantum Optics", ed. by R.J. Glauber (Academic, New York 1969).

5. F.T. Arecchi, R. Meucci, and W. Gadomski, Phys Rev. Lett. **58**, 2205 (1987).

6. L.P. Shil'nikov, Dokl. Akad, Nauk SSSR **160**, 558 (1965); L.P. Shil'nikov, Mat. Sb. **77**, 119, 461 (1968); **81**, 92, 1213 (1970).

7. P. Glendinning and C. Sparrow, J. Stat. Phys. **35**, 645 (1984); P. Gaspard, R. Kapral and G. Nicolis, J. Stat. Phys. **35**, 697 (1984).

8. F. Argoul, A. Arneodo, and P. Richetti, Phys. Lett. A **120**, 269 (1987).

9. F.T. Arecchi, A. Lapucci, R. Meucci, J.A. Roversi and P. Coullet, Europhys. Lett. **6**, 677 (1988).

10. F.T. Arecchi, W. Gadomski and R. Meucci, Phys. Rev. A **34**, 1617 (1986).

11. A. Arneodo, P.H. Coullet, E.A. Spiegel and C. Tresser, Physica **14D**, 327 (1985).

NORMAL FORMS FOR SYMPLECTIC MAPS AND STABILITY OF BEAMS IN PARTICLE ACCELERATORS

by

G. Turchetti †

and

A. Bazzani, M. Giovannozzi, G.Servizi†, E. Todesco

Dipartimento di Fisica, Università di Bologna, ITALY
INFN, Sezione di Bologna, ITALY

Abstract

The analysis of the nonlinear effects on the stability of the betatronic motion has become quite important after the introduction of very high field superconducting dipoles. Indeed the multipolar errors are not avoidable and the properties of the single particle motion in presence of highly nonlinear fields must be understood. Technics alternative to tracking are necessary in order to compute the relevant nonlinear parameters and carry out correction strategies. We present a description of the magnetic lattice in terms of tranfer maps both in the linear and in the nonlinear case. For nonresonant linear tunes the use of the Birkhoff normal forms turns out to be the analogue of the discrete version of the Courant Sneyder theory; in the resonant case the extension is more subtle due to the more intricate topology of the orbits. After reviewing briefly the theory of the non resonant and resonant normal forms and of the interpolating hamiltonians, we examine the convergence properties of the Birkhoff series. Heuristic and rigorous arguments are provided to describe their asymptotic nature and the consequences of a Nekhoroshev like stability theorem are illustrated. It is also shown how informations on the dynamic aperture are obtained from the divergence of the series. The relation of the dynamic aperture with the stable and unstable manifolds of the unstable fixed points of the maps, computable with hyperbolic normal forms, is discussed. Finally the results on the tune shift and smear for the CERN large hadron collider (LHC) for the uncopuled vertical and coupled vertical and horizontal betatronic oscillations are reviewd. The results on the SPS 1989 experiment on the tune shift and the dynamic aperture are compared with the normal form results which turn out be very satisfactory. Finally the correction strategy on the multipolar errors is described. The use of the normal forms allows a unique solution to be found as a result of the minimization of the tune shift; moreover in the symmetric case the correction parameters are obtained by solving linear systems and the method turns out to be by far more efficient and reliable with respect to any other method based on tracking techniques.

---

† Work supported in part with a MPI 60% grant

Introduction

The analysis of the betatronic motion in the new accelerators, where high field superconducting magnets are used, cannot be performed within the classical theory developed by Courant and Snyder [1]. Indeed such a theory applies only when the bending magnets are pure dipoles and the focusing-defocusing lenses pure quadrupoles so that the transverse motion of the charged particles is described by the linear Hill's equations.

In the accelerators of the present generation the only nonlinearities are given by weak sextupole fields introduced to correct the chromatic effects. Even though a complete theoretical understanding on the stability of the beam is lacking, the numerical simulations prove to be adequate for the machine design.
The new high field dipoles and quadrupoles have so important multipolar errors [2] (higher order unavoidable multipoles) that accurate correction strategies must be found in order to reduce the non linear effects and to bring the tune shift, the short time escape region (dynamic aperture) and the long time diffusion to acceptable values. Furthermore the effect of the random errors in the multipoles due to construction imperfections and to the randomness in the current rise should be considered.
An approach to these problems based on simulations is certainly not satisfactory for many reasons: first of all the tracking codes are quite slow and dynamic parameters such as the tune shift cannot always be accuratly computed. The correction strategies, which require minimizations on high dimensional parameter spaces, can be performed by tracking only using very empirical recipies of limited reliability. Moreover even the analysis of the phase space portraits of the orbits is a non trivial problem and is usually solved by considering only the projections onto the coordinate planes (x, p_x) and (y, p_y) [3]. The usual analytic approach based on the first or second order perturbation theory applied to the Hamiltonian which describes the magnetic lattice is adequate only to produce very qualitative results [4]. Indeed we should not forget that an accelerator is a discontinuous system (the fringe effects in the magnets are actually negligible). As a consequence each element of the lattice (magnet or drift) is described by a tranfer map and the sector of the accelerator corresponding to a superperiod is described by a map obtained by composing the transfer maps of all its elements. If each transfer map is a polynomial symplectic map, the superperiod map M will also be symplectic, but given by a polynomial of incredibly high order (if there are m magnetic elements described by polynomial maps of degree d, the degree of the superperiod map will be d^m). A truncation $[M]_{\leq N}$ to a polynomial of low order N, which can be recursively computed with manageble algorithms, cannot be used to track at high orders, since it is no longer exactly symplectic [5] (local expanding or contracting behaviours become very evident by increasing the number of iterations). The Lie algebraic tecniques, considered by A. Dragt [6], allow to replace the map $[M]_{\leq N}$ with a symplectic one of the form $\exp\{[H]_{\leq N}\}$; this map can be easily iterated, however since no control on the error due to the symplectification of the map $[M]_{\leq N}$ is carried out, the method is not reliable.

A solution to this problem is given by the normal forms introduced by Birkhoff more than fifty years ago [7]. The method allows to conjugate the orbits of a symplectic map whose linear frequencies are non resonant with orbits on tori and to compute the winding numbers associated to each orbit. It is known that the series defining the normal forms are asymptotic becouse there cannot be tori with rational frequencies, according to the Poincaré theorem and the KAM theory[8]. However a careful analysis of the asymptotic properties of these series was carried only very recently, first by Henon[9] and then by the Bologna group, which explained the divergence mechanism suggested by the results obtained at very high orders with an algorithmic approach[10,11]. The same group extended the theory of normal forms to the symplectic maps of $\mathbf{R}^{2d}$[12,13] both in the non resonant and resonant case, where the interpolating hamiltonians were explicitly introduced[14] and obtained rigorous estimates on the remainders of the Birkhoff series [15,16]. As a consequence, the theorem which bounds the Arnold diffusion, first proved by Nekhoroshev for the flows was extended to the symplectic maps [16].

The first applications of the Birkhoff normal forms to beam dynamics were carried out by the Bologna group and concerned the computation of tune shift and smear for the beam-beam interaction[17] and the vertical betatronic motion in a LHC lattice[18]. A realistic analysis of the full betatronic motion of LHC[19] was carried out simultaneously as the Berkeley group for SSC[20].

Very recently the SPS experiment with strong sextupoles was succesfully analyzed[21] and a very good strategy of basically analytical nature was carried out for the multipolar corrections of LHC[22].

At present the challenging problems are the following:

1) to understand the origin dynamical aperture and its connection with the validity domain of the Birkhoff series
2) to compare the theoretical estimates for the remainders and the Arnold diffusion with numerical experiments
3) to provide a hyerarchiral structure for the multi-dimensional resonances and to describe the geometrical structure of the resonant orbits
3) to consider the faster modulation diffusion in order to explain the beam lifetime in the recent experiments
4) to develop a theoretical understanding of the stochastic perturbations to a symplectic map

In the first two items some progress is being made and will be reported here. However a rigorous mathematical treatment involves the study of Julia sets for maps of $\mathbf{C}^2$ and $\mathbf{C}^4$, a really formidable task. As far as the applications are concerned a lot remains to be done, both for large and small lattices.

The plan of the work is the following:

In section 2 we describe an accelerator lattice using the transfer map language.

In section 3 we describe a simple model of a FODO cell which reduces to the Henon Map and we consider the normal forms for maps of the plane in the non resonant and and in the resonant case.

In section 4 we consider the problem of divergence of the Birkhoff series, the dynamic aperture and outline the analytic results and the estimates on the Arnold Diffusion.

In section 5 we consider three relevant applications to accelerator physics: the LHC uncorrected lattice parameters, the SPS 1988 experiment and finally the correction of the multipolar errors in LHC.

Transfer map description of a lattice

In a circular accelerator we consider the reference circular orbit and choose an orthogonal reference plane where the z axis is tangent to the orbit, x is the other horizontal axis and y is the vertical axis, see figure 1.

Fig.1

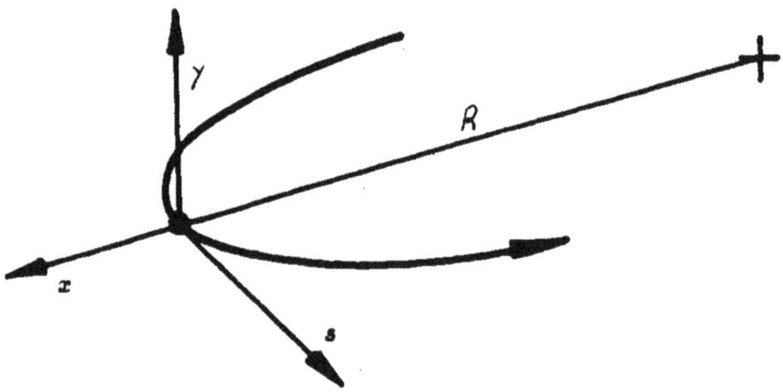

Reference system.

Letting R be the radius of the orbit, v, p the particle velocity and momentum and $s = vt = R\theta$ the arc length, the equations of motion in the x, y plane read

$$\begin{aligned} \frac{d^2x}{ds^2} &= \frac{1}{R^2} + \frac{e}{pc}B_y \\ \frac{d^2y}{ds^2} &= -\frac{e}{pc}B_x \end{aligned} \tag{2.1}$$

Letting $w = x + iy$ the multipolar expansion of the magnetic field $\mathbf{B}$ reads

$$B_y + iB_x = \sum_{n\geq 1} \frac{B_n}{n!} w^n \tag{2.2}$$

where B_n are real coefficients. Denoting with B_0 the vertical dipole field such that $pc = eB_0R$ we set

$$k_j = \frac{B_j}{B_0 R} \qquad j \geq 1 \tag{2.3}$$

and the equations of motion read

$$\begin{aligned} \frac{d^2x}{ds^2} + \left(\frac{1}{R^2} - k_1\right)x &= \frac{k_2}{2}(x^2 - y^2) + \frac{k_3}{3!}(x^3 - 3xy^2) + \ldots \\ \frac{d^2y}{ds^2} + k_1 y &= -k_2 xy - \frac{k_3}{3!}(3x^2y - y^3) + \ldots \end{aligned} \tag{2.4}$$

To simplify we assume that B_0 is constant, namely that in absence of quadrupoles and higher multipoles the orbit is perfectly circular, while the functions $k_j(s)$ are discontinuous: their value has a step discontinuity as one passes from a magnet to onother. If the machine is linear, namely $k_2 = k_3 = \ldots = 0$ the equations decouple and the solution can be written

$$\begin{aligned} x(s) &= A_x\sqrt{\beta_x(s)}\sin(\psi_x(s) + \delta_x) \\ y(s) &= A_y\sqrt{\beta_y(s)}\sin(\psi_y(s) + \delta_y) \end{aligned} \tag{2.5}$$

where $A_{x,y}$ and $\delta_{x,y}$ are constants and

$$\psi_x(s) = \int_0^s \frac{d\sigma}{\beta_x(\sigma)}, \qquad \psi_y(s) = \int_0^s \frac{d\sigma}{\beta_y(\sigma)} \tag{2.6}$$

and the periodicity imposes $\psi_{x,y}(0) = \psi_{x,y}(\ell)$ where ℓ is the arc length of a superperiod.

Let the superperiod be formed by n elemets such that the first element is in the arc $[0, s_1]$, the second in $[s_1, s_2]$ and the last in $[s_{n-1}, \ell]$ and denote by $\mathbf{L}_1, \mathbf{L}_2, \ldots, \mathbf{L}_n$ the corresponding linear part of the transfer maps $\mathbf{M}_1, \mathbf{M}_2, \ldots, \mathbf{M}_n$, we recall that denoting with $\mathbf{x} = (x, p_x, y, p_y) \in \mathbf{R}^4$, where $p_x = dx/ds$ and $p_y = dy/ds$ the solution (2.6) can be written

$$\mathbf{x}(s) = \mathbf{T}(s)\mathbf{R}(s)\mathbf{T}^{-1}(0)\mathbf{x}(0), \qquad \mathbf{T}(0) = \mathbf{T}(\ell) \tag{2.7}$$

where

$$\begin{aligned} \mathbf{T}(s) &= \begin{pmatrix} T_x(s) & 0 \\ 0 & T_y(s) \end{pmatrix}, \quad \mathbf{R}(s) = \begin{pmatrix} R(\psi_x(s)) & 0 \\ 0 & R(\psi_y(s)) \end{pmatrix} \\ T_{x,y} &= \begin{pmatrix} \beta_{x,y}^{1/2} & 0 \\ \gamma_{x,y} & \beta_{x,y}^{-1/2} \end{pmatrix} \qquad R(\alpha) = \begin{pmatrix} \cos\alpha & -\sin\alpha \\ \sin\alpha & \cos\alpha \end{pmatrix} \end{aligned} \tag{2.8}$$

and $\gamma_{x,y} = \dot{\beta}_{x,y}(s)/2\sqrt{\beta_{x,y}(s)}$. The matrix $\mathbf{L}(s)$ for Poincaré map at the point s is defined by

$$\mathbf{x}(s + \ell) = \mathbf{L}(s)\mathbf{x}(s) \qquad \mathbf{L}(s) = T(s)\mathbf{R}(\ell)\mathbf{T}^{-1}(s). \tag{2.9}$$

Indeed using (2.7) and taking the periodicity of $\mathbf{T}(s)$ into account one has

$$\mathbf{x}(s+\ell) = \mathbf{T}(s+\ell)\mathbf{R}(s+\ell)\mathbf{T}^{-1}(0)\mathbf{x}(0) = \mathbf{T}(s)\mathbf{R}(s+\ell)\mathbf{R}(-s)\mathbf{T}^{-1}(s)\mathbf{x}(s). \tag{2.10}$$

The discrete description of the lattice is extremely simple since is based on the transition matrices $\mathbf{L}_j$ corresponding to the single elements of the superperiod where $k_1(s)$ is constant. According to figure 2 we have

$$\mathbf{L}(0) = \mathbf{L}_n\mathbf{L}_{n-1}\ldots\mathbf{L}_1 = \mathbf{T}(0)\mathbf{R}(\ell)\mathbf{T}(0)^{-1} \tag{2.11}$$

and consequently

$$\mathbf{L}(s_k) = \mathbf{L}_k\mathbf{L}(s_{k-1})\mathbf{L}_k^{-1} \tag{2.12}$$

so that comparing with (2.9) evaluated at $s = s_k$ we obtain

$$\mathbf{T}(s_k) = \mathbf{L}_k\mathbf{T}(s_{k-1}). \tag{2.13}$$

It is therefore evident that once we have computed $\mathbf{L}(0)$ and from it $\mathbf{T}(0)$, all the matrices corresponding to different Poincaré sections of the lattice are given by the last recurrence.

Fig.2

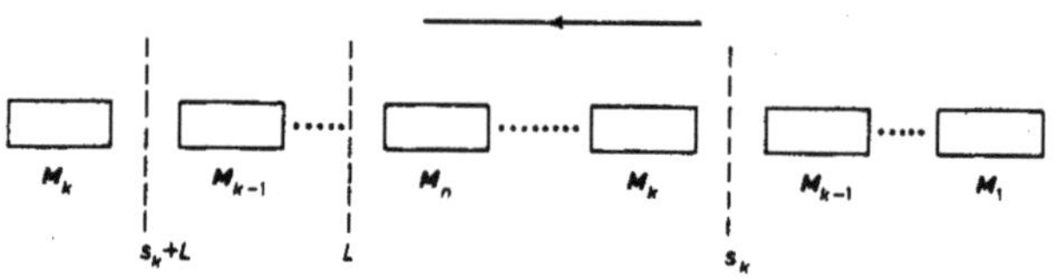

Outline of the lattice.

The matrices $\mathbf{T}(s_k)$ are all uniquely defined up to a rotation. Without any loss of generality we can fix this rotation to be the identity. The matrix $\mathbf{R}(\ell)$ gives the tunes $\nu_{x,y} = \psi_{x,y}(L)/2\pi$ while $\mathbf{T}(s_k)$ give the geometric deformation factors. In any section $s = s_k$, the matrix $T(s_k)$ defines the change to a normalized set of coordinates

$$\hat{\mathbf{x}} = \mathbf{T}^{-1}(s_k)\mathbf{x} \tag{2.14}$$

such that $\hat{x}^2 + \hat{p}_x^2$ and $\hat{y}^2 + \hat{p}_y^2$ are invariants of the motion. The geometrical meaning is also evident, we tranform an orbit whose (x, p_x) and (y, p_y) projections are ellipses into an orbit whose projections are circles. However the transformation depends on the section we are considering whereas the tunes (which are the true invariants) do not depend.
From now on we shall choose our refence section ar $s = 0$.

When the sextupoles and higher multipoles are present the transition matrices $\mathbf{M}_j$ are given by the solution of the Cauchy problem in the interval $[s_{j-1}, s_j]$ of the autonomous differential equations (2.4) (the $k_1(s), k_2(s), \ldots$ are constant in this interval) , with polynomial nonlinearities. Symplectic approximations to $\mathbf{M}_j$ are obtained with the kick approximation which corresponds to inserting $(s_k - s_{k-1})\ \delta(s - \frac{(s_k - s_{k-1})}{2}$ in the r.h.s. of eq. (2.4). In this approximation the map $\mathbf{M}_j$ is a polynomial whose degree is the same as for the non linear terms in (2.4).The scheme can be refined by dividing $[s_{k-1}, s_k]$ into ℓ equal subintervals, computing for each of them the transition map in the kick approximation and finally expressing $\mathbf{M}_j$ by composing all of them. For instance the transfer map $\mathbf{F}_j$ corresponding to a dipole or a quadrupole will be given by

$$\begin{pmatrix} x' \\ p'_x \\ y' \\ p'_y \end{pmatrix} = \mathbf{L}_j \begin{pmatrix} x \\ p_x + \frac{k_2}{2}\ell_j(x^2 - y^2) + \frac{k_3}{3!}\ell_j(x^3 - 3xy^2) + \ldots \\ y \\ p_y - k_2\ell_j\, xy - \frac{k_3}{3!}\ell_j(3x^2y - y^3) + \ldots \end{pmatrix} \tag{2.15}$$

where $\ell_j = s_j - s_{j-1}$ the length of the element.
The full map on a superperiod is obtained by composing all the maps for the single elements and reads

$$\mathbf{F} = \mathbf{M}_n \circ \mathbf{M}_{n-1} \circ \ldots \circ \mathbf{M}_1 \tag{2.16}$$

If the $\mathbf{M}_j$ are symplectic then $\mathbf{F}$ is symplectic too. We observe that denoting with $[\mathbf{F}]_{\leq N}$ the truncation of $\mathbf{F}$ to a polynomial of order N the algorithm to compute it is obvious and reads

$$\mathcal{F}_j = [\mathbf{M}_j \circ \mathcal{F}_{j-1}]_{\leq N} \quad j = 1, \ldots, n \qquad \mathcal{F}_0 = I \quad \mathbf{F} = \mathcal{F}_n. \tag{2.17}$$

The algorithmic approach allows to reach significantly high orders namely $N \geq 10$ for maps of $\mathbf{R}^4$ and $N \geq 30$ for maps of $\mathbf{R}^2$. The use algebraic manipulators is not suggested becouse they are much slower and typically do not allow to overcome $N = 5$.

Normal forms

For the linear part $\mathbf{L}$ of the Poincaré map we have introduced the normalized coordinates $\hat{\mathbf{x}}$ through (2.14). It is then convenient to use complex coordinates

$$z = (z_1, z_2) \in \mathbf{C}^2 \qquad z_1 = \hat{x} + i\hat{p}_x, \quad z_2 = \hat{y} + i\hat{p}_y \tag{2.18}$$

so that linear part of the map can be written as

$$z' = e^{i\omega} z \qquad z = (z_1, z_2) \in \mathbf{C}^2, \quad e^{i\omega} = \mathrm{diag}(e^{i\omega_x}, e^{i\omega_y}) \tag{2.19}$$

together with its complex conjugate. As a consequence the normal form of the map in the linear case is given by a direct product of rotations. The nonlinear map will read

$$z' = F(z, z^*) = e^{i\omega} z + [F]_2 + \ldots + [F]_n + \ldots \tag{2.20}$$

where $[\ ]_n$ denotes the projection on the homogeneous polynomial in z, z^* of order n.
In the general case we say that a map $U : \mathbf{C}^2 \times \mathbf{C}^2 \to \mathbf{C}^2$ is in *normal form* with respect to $\omega \in \mathbf{R}^2$ if it commutes with the group generated by $e^{i\omega}$ namely. This is the full group of

the rotations on the T^2 torus if ω is non resonant, conversely it is a continuous subgroup of dimension 1 or a discrete group if ω satisfies a resonance condition of order 1 or 2 respectively *. As a consequence any map is in normal form if

$$\Delta U \equiv U(e^{i\omega}z, e^{-i\omega}z^*) - e^{i\omega}U(z,z^*) = 0. \tag{2.21}$$

If the map G is symplectic and ω non-resonant namely $(\nu, 1)\cdot \mathbf{k} = 0,\ \mathbf{k} \in \mathbf{Z}^3 \Rightarrow \mathbf{k} = 0$ then the map takes a very simple form

$$U(z,z^*) = e^{i\Omega(zz^*)}z \qquad \Omega(zz^*) \equiv \Omega(z_1z_1^*, z_2z_2^*) \tag{2.22}$$

where $\Omega(\rho_1, \rho_2)$ is a real function such that $\partial\Omega_l/\partial\rho_j = \partial\Omega_j/\partial\rho_l$.
The map (2.22) is a rotation on a torus with the rotation angles depending on the invariants $z_1z_1^*$, $z_2z_2^*$ and naturally provides the *interpolating flow* $\exp(i\Omega t)z$ for its orbits.
This is no longer the case when the map is resonant: in this case the iteration of the map is no longer trivial. Indeed in this case the interpolation of the orbits require the construction of an interpolating hamiltonian, which will be integrable only in the case of a resonance of order 1 (in this case the hamiltonian splits into a pendulum like plus an integrable hamiltonians). The construction of this interpolation, achieved with series whose existence is proved, is not straightforward. [11]

If a given map F is not in normal form we look for a change of variables $z \to \varsigma$ which brings it to the normal form (2.22). Letting

$$z = \Phi(\varsigma,\varsigma^*) = \varsigma + \sum_{n\geq 2}[\Phi]_n(\varsigma,\varsigma^*) \tag{2.23}$$

be a symplectic diffeomorphism close to the identity the following functional equation must be satisfied:

$$F\circ\Phi - \Phi\circ U = 0, \tag{2.24}$$

where we have used the notation $F\circ\Phi = F(\Phi(\varsigma,\varsigma^*), \Phi^*(\varsigma,\varsigma^*))$ and similarly for the second term. It is a well-known consequence of the KAM theory that, even for a diophantine ω, equation (2.24) cannot have an analytic solution. As a consequence denoting with $[\Phi]_{\leq N}$ and $[\Omega]_{\leq N}$ the Taylor series of Φ and U truncated at order N, we replace equation (2.24) with

$$F\circ[\Phi]_{\leq N} - [\Phi]_{\leq N}\circ[U]_{\leq N} = R_N \tag{2.25}$$

where R_N is a remainder given by a series which begins at order $N+1$. Letting $[S]_n$ be the homogeneous component of order n in the l.h.s. of equation (2.25) once we subtract the order n unknowns $[U]_n$ and $\Delta[\Phi]_n$, the recurrent equations are obtained by projecting (2.25) into the space of normal form polynomials and its complement. Letting Π be the projector into the space of normal form polynomials we have

$$\begin{aligned} [U]_n &= \Pi\,[S]_n \\ \Delta(1-\Pi)[\Phi]_n &= (1-\Pi)[S]_n. \end{aligned} \tag{2.26}$$

* We say that $\omega\in\mathbf{R}^2$ is resonant if $(\nu,1)\cdot\mathbf{k}=0$ for $\forall\mathbf{k}\in\mathcal{M}\subset\mathbf{Z}^3$ and the order of the resonance is the dimension of the subset of $\mathbf{Z}^3$

It has been proved [13], using a symplectic generator, that equations (2.25) admit solutions $[\Phi]_n$ and $[U]_n$ symplectic, up to order $n \leq N$. In the non resonant case we have $[U]_n = [\varsigma \exp(i[\Omega]_{\leq N})]_n$ and a consequence a real frequency function $[\Omega]_{\leq N}$ is found. The solution $[\Phi]_{\leq N}$ is not unique but depends on a gauge $e^{i\Xi(\varsigma\varsigma^*)}$ where Ξ is an arbitrary function [16]. We stress that in this problem the natural expansion parameter is

$$\|z\| = \max_j |z_j| \tag{2.27}$$

and consequently we expand any function in a series of homogeneous polynomials in z and z^*. Letting $\Psi = [\Phi]^{-1}_{\leq N}$ define the inverse transformation we introduce the remainder $\hat{R}_n(t)$ corresponding to (2.25) iterated $t \in \mathbf{Z}$ times according to

$$\Psi \circ F^t \circ [\Phi]_{\leq N} = e^{it[\Omega]_{\leq N}}\varsigma + \hat{R}_N(t). \tag{2.28}$$

Area preserving maps

In order to illustrate the normal forms and their properties we consider the simpler, but far from trivial case of symplectic maps of $\mathbb{R}^2$ area *preserving maps*. These maps describe the betatronic motions in the limiting case in which the coupling between horizontal and vertical motion can be neglected. We shall consider a very special model, the Henon quadratic map to which the most general quadratic symplectic map with an elliptic fixed point can be reduced with a linear transformation. This map depends on no parameter but the frequency of the linear part. It describes a very simple lattice whose superperiod is a single FODO given by a focusing quadrupole a drift a defocusing quadrupole and a dipole with a multipupolar error treated in the sigle kick approximation. The map $\mathbf{F}$ is then given by

$$\begin{pmatrix} x' \\ p'_x \end{pmatrix} = L_x \begin{pmatrix} x \\ p_x + \frac{k_2}{2}\ell x^2 + \frac{k_3}{3!}\ell x^3 \end{pmatrix} \tag{3.1}$$

where ℓ is the length of the non linear magnets.
Letting β_x, γ_x be the constants associated to the symplectic matrix L_x according to $L_x = T_x R(2\pi\nu_x) T_x^{-1}$ where T_x is given by (2.8), the following change to the linear normalized coordinates $(\hat{x}, \hat{p}_x)$

$$x = \sqrt{\beta_x}\hat{x}, \quad p_x = \gamma_x\hat{x} + \frac{1}{\sqrt{\beta_x}}\hat{p}_x$$

followed by a scaling and the usual introduction of complex coordinates

$$z = \frac{k_2}{2}\ell\beta_x^{3/2}(\hat{x} + i\hat{p}_x)$$

leads to the map

$$z' = e^{-i\omega}\left[z + \frac{i}{4}(z+z^*)^2 + \frac{i\kappa_3}{8}(z+z^*)^3 + \ldots\right] \tag{3.2}$$

where $\kappa_3 = 4k_3/(3k_2^2\ell^2\beta)$ and $\omega = 2\pi\nu_x$. If the multipolar errors all vanish except for the sextupoles we have the quadratic Henon map, which depends only on the frequecy ω. From now on we shall restrict all our considerations to this model.

Normal forms

We look for the order N truncation of a symplectic diffeomorphism $\Phi : \varsigma \to z$ where ς is the new coordinate where the map takes *normal* form U. One obtains the functional equation (2.24) for Φ and U from the formally commutative diagram

$$\begin{array}{ccc} (z,z^*) & \xrightarrow{F} & (z',z^{*\prime}) \\ \phi \uparrow & & \uparrow \phi \\ (\varsigma,\varsigma^*) & \xrightarrow{U} & (\varsigma',\varsigma^{*\prime}) \end{array} \tag{3.3}$$

The functional equation is solved by the recurrence (2.26) where one has

$$\Phi(\varsigma,\varsigma^*) = \varsigma + \sum_{n\geq 2}\left[\sum_{k=0}^{n} a_{n,k}\varsigma^k\varsigma^{*n-k}\right] \tag{3.4}$$

and in the non resonant case, namely when $\omega/2\pi$ is irrational,

$$U(\varsigma,\varsigma^*) = e^{i\left[\omega+\sum_{n\geq 1}(\varsigma\varsigma^*)^n\Omega_{2n}\right]}\varsigma \tag{3.5}$$

For $n \to \infty$ the series defining Φ and Ω cannot converge; indeed between KAM curves there is a dense set of chains of islands of period q_i, which occur whenever the winding number is rational, that is

$$\frac{\Omega(\varsigma\varsigma^*)}{2\pi} = \frac{p_i}{q_i} \tag{3.6}$$

This complicate structure of the fase space is the leading reason of the divergence of perturbative series: the normal form is 'too simple' to reproduce the dynamics of the real map.

Resonant normal forms

With resonant normal forms we conjugate the map F with a map U whose space structure is the following: KAM tori separated by chains of islands whose periods p_i are multiple of a number q which characterizes the leading resonance according to

$$\frac{\omega}{2\pi} - \frac{p}{q} = \epsilon \quad \epsilon \ll 1 \tag{3.7}$$

In this way we manage to reproduce the dynamics of the original map f very well also close to the resonance when the standard non resonant approach fails (fig. 3).

Fig.3

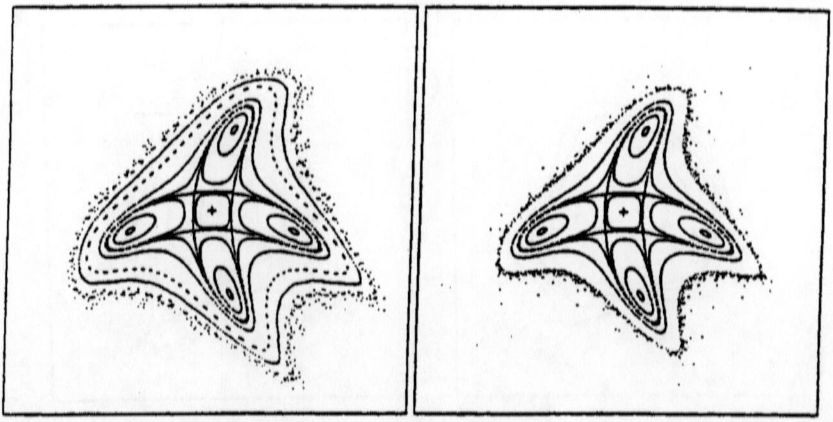

Phase portrait of Henon's map (left) and its resonant normal form for a resonance of order 4.

The structure of the resonant normal form is obtained by imposing the simmetry with the group of discrete rotations of an angle

$$\omega_R = 2\pi \frac{p}{q}$$

$$u(e^{i\omega_R}\varsigma, e^{-i\omega_R}\varsigma^*) = e^{i\omega_R} u(\varsigma, \varsigma^*) \tag{3.8}$$

that is

$$u(\varsigma, \varsigma^*) = e^{i\omega_R + \epsilon} \left[\varsigma + \sum_{s,l} (\varsigma\varsigma^*)^s (u_{s,l}\varsigma^{lq+1} + u_{s,-l}\varsigma^{*lq-1}) \right]. \tag{3.9}$$

The perturbative series will not converge anyway: infact arbitrarily close to the origin we have chains of islands whose period q' is not a multiple of q. The conjugation function at order $q' - 1$, trying to conjugate diffeomorphically two structures, that are topologically different will lose invertibility (fig. 4), shrinking the radius of convergence of the series. For $N \to \infty$ this radius will go to zero. This picture gives also a rough idea about the limit of validity of this perturbative approach: besides a certain order $\overline{N}$ depending on $|\varsigma\varsigma^*|$, the normal form will not reproduce the dynamics of the map f.

While the non resonant form is area preserving, the resonant one as written in (2.7) is area preserving only up to order $N - 1$ (fig. 5).

Nevertheles it is possible to construct an interpolating hamiltonian (fig. 6) of the normal form which satisfies

$$H(\varsigma, \varsigma^*) = H(e^{i\omega_R}\varsigma, e^{-i\omega_R}\varsigma^*) \tag{3.10}$$

Fig.4

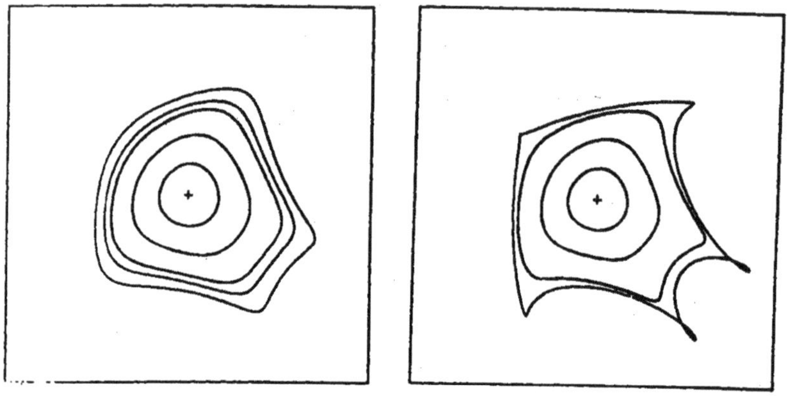

Loss of invertibility of the conjugation function due to a resonance of order 5.

Fig.5

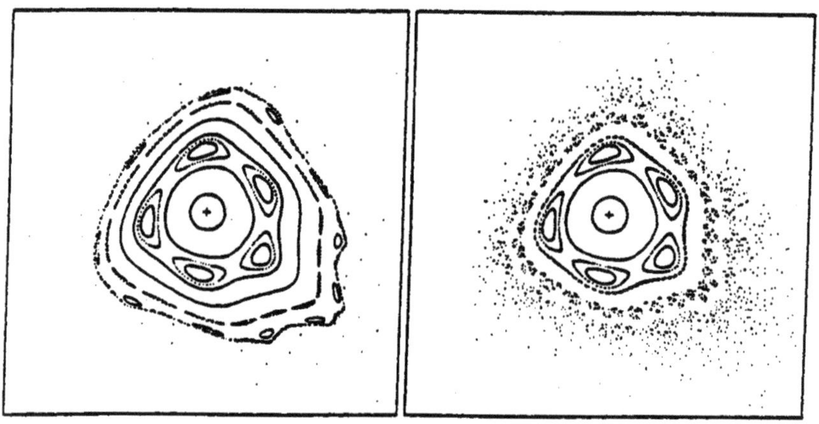

Non conservative behaviour of the resonant normal form (right) for the Henon's map (left).

The structure of this hamiltonian letting $\varsigma = \rho e^{i\theta}$ is

$$H(\varrho,\vartheta) = \epsilon(\varrho^2) + \sum_{s,l} h_s^l (\varrho^2)^{s+\frac{lq}{2}} \cos(lq\vartheta + \varphi_l). \qquad (3.11)$$

We recall that the action variable is given by J where $\rho = \sqrt{2J}$.

Stability analysis

Fig.6

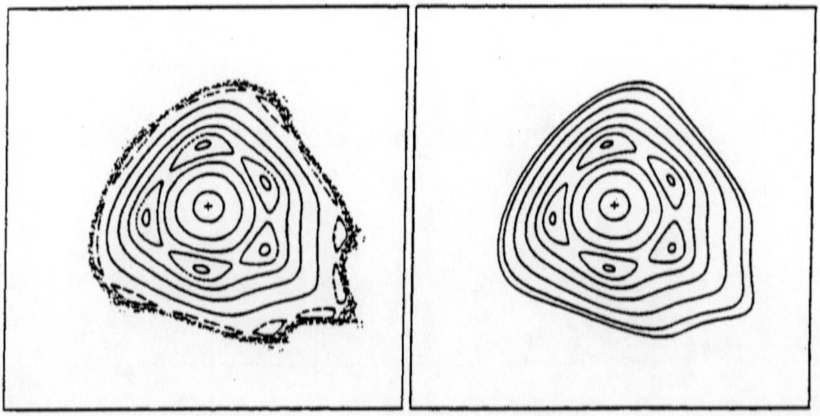

Interpolating Hamiltonian at order 7 (right) of the Henon's map (left) near a resonance of order 5.

The study of interpolating hamiltonian allows to check the stability of the fixed point (0,0). This analysis is not trivial since in some cases the stability of the linearized map is not sufficient to insure the stability of the map itself. To solve this problem one considers the level curve passing through the origin. If the curve reduces to the origin itself, then the origin is stable; otherwise the curve is necessarily a separatrix and the origin is unstable. The following results hold:

i)

$$\frac{\omega}{2\pi} = \frac{p}{3}$$

$$H = h_0^1 \varrho^3 \cos 3\vartheta + h_2^0 \varrho^4 \tag{3.12}$$

(It's not restrictive assume $h_2^0 \neq 0$) The separatrix equation has solution:

$$\varrho = \frac{h_0^1}{h_2^0} \cos 3\vartheta \tag{3.13}$$

the point is unstable provided that $h_0^1 \neq 0$ (as for instance for Henon's map).

ii)

$$\frac{\omega}{2\pi} = \frac{p}{4}$$

$$H = (h_0^1 \cos 4\vartheta + h_2^0)\varrho^4 + \varrho^6 (h_3^0 + h_1^1 \cos 4\vartheta) \tag{3.14}$$

$$\varrho^2 = \frac{h_0^1 \cos 4\vartheta + h_2^0}{h_3^0 + h_1^1 \cos 4\vartheta} \tag{3.15}$$

Here depending on the values of h_0^1, h_2^0, h_3^0, h_1^1 one can have stability or instability. Henon's map in this case exhibits a singular behaviour. Since $h_0^1 = -h_2^0$, if we use perturbative

Fig.7

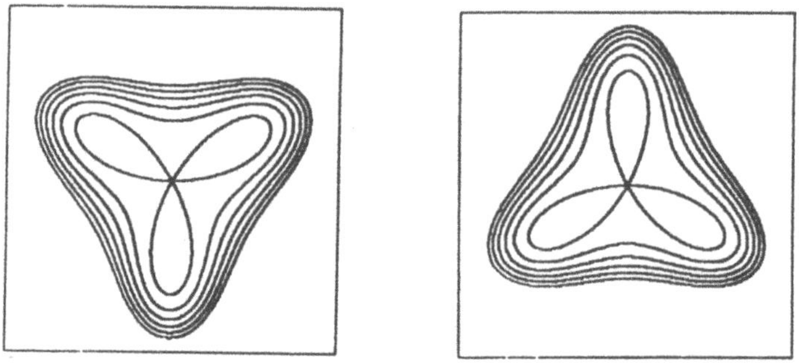

Two cases of a resonance of order 3 for the Henon's map.

theory only up to order 3 (corresponding to order 4 in the interpolating hamiltonian, namely up to $(\varrho)^2$), then the origin appears to be unstable while for $N \geq 4$ it is stable, as one can observe in the phase space portrait obtained from tracking. Moreover, when we approach the origin, the orbits do not look like circles, but rather like hyperbolae, even though they join smoothly at a finite distance from the origin, see (fig. 8). This singular phase space structure disappears if we add a small pertubation to the map:

$$f_\epsilon(z) = e^{i\omega_R}\left[z + i(z+z^*)^2 + i\epsilon(z+z^*)^3\right] \tag{3.16}$$

iii)

$$\frac{\omega}{2\pi} = \frac{p}{q} \quad q \geq 5$$

For these resonances the origin is always stable, provided that exist a $h_i^0 \neq 0$ of order less than the first resonant term, that is the most common case. Indeed, it is possible to build a singular map in the form

$$f_s(z) = e^{i\omega}\left[z + i(z+z^*)^2 + i\alpha_3(z+z^*)^3 + i\alpha_5(z+z^*)^5 + \ldots\right] \tag{3.17}$$

choosing the coefficients α_i to set the coefficients of the normal form $h_i^0 = 0$ up to a certain order. In this way all the resonances up to the same order are unstable. This procedure is used in accelerators physics, when we insert non linear magnets (sextupoles, octupoles....) to minimize the tune shift

$$\delta\nu = \Omega_2(\varsigma\varsigma^*) + \Omega_4(\varsigma\varsigma^*)^2 + \ldots \tag{3.18}$$

of the transfer map on a superperiod.Infact the strenghts of n-poles are proportional to α_i, and the same relationship holds between Ω_i and h_i. That is, when one tries to cancel the

Fig.8

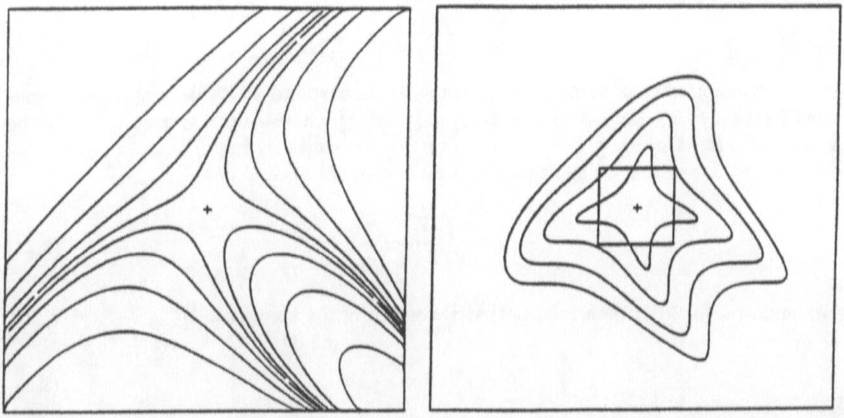

Phase space of the interpolating Hamiltonian for a resonance of order 4 computed up to 4(left) and 6(right) order.

coefficients of the tune shift the corresponding resonance becomes unstable. Finally, the following scaling law for the width of islands created by a nearby resonance (3.7) holds:

$$\Delta \propto \epsilon^{\frac{q}{4}} \qquad (3.19)$$

Tune shift and smear

The nonresonant and resonant normal forms naturally provide information on the variatione of the frequency with the amplitude (tune shift) and the deformation of the orbits. It is interesting to compare the results obtained from a direct eveluation of the orbit (*tracking*) with the normal form results. We recall that given a map $F : \mathbf{R}^2 \to \mathbf{R}^2$, $\mathbf{x}_j = F^{\circ j}\mathbf{x}$ with $j = 0, 1, \ldots$ be the points of its orbits and α_j be the phase advance at step j namely the angle between the vectors $\mathbf{x}_{j-1}$ and $\mathbf{x}_j$, then the tune is defined by

$$\nu(\mathbf{x}) = \frac{1}{2\pi} \lim_{n \to \infty} \frac{1}{n} \sum_{j=1}^{n} \alpha_j \qquad (3.20)$$

The existence of such a limit was first proved by Poincaré for the circle maps.The limit will exist whenever the orbit passing through $\mathbf{x}$ is a KAM curve, a fixed point of $F^{\circ m}$ or a chain of islands. As a consequence if we let $\mathbf{x}$ vary on the the x axis the tune shift will be a smooth function of x only on the complement of a Cantor set. Nevertheless what is observed is a continuous curve with plateaus corresponding to the crossing of islands where the tune remains constant. Numerically only a very small small number of these plateaus is observed for a very simple reason: the width of a resonance p/q vanishes as $1/\sqrt{(q^q)}$.

Indeed if we consider the lowest order perturbation theory we have that the condition for a resonant tune reads

$$\nu = \frac{\Omega}{2\pi} \simeq \frac{\omega}{2\pi} + \Omega_2 \rho^2 = \frac{p}{q} \tag{3.21}$$

Now the resonances that best approximate $\omega/2\pi$ correspond to its development in contunued fraction p_j/q_j :indeed $\omega/2\pi - p_j/q_j \simeq (-1)^j/q_j^2$ whereas for a given q typically the best approximation p/q to $\omega 2/\pi$ is $\sim 1/q$. As a consequence letting $\epsilon_j = \omega/2\pi - p_j/q_j$ for j even or odd depending on the sign of Ω_2 the resonance occurs for

$$\rho \simeq \sqrt{\frac{-\epsilon_j}{\Omega_2}} \simeq \frac{1}{q_j} \tag{3.22}$$

From equation (3.19) it then follows that the width of the resonance is

$$\Delta \propto \epsilon_j^{q_j/4} \propto \frac{1}{\sqrt{q_j^{q_j}}} \tag{3.32}$$

In a disk of radius r the area of these chains of islands consequently is certainly smaller than $r^{1/2r}$ and vanishes faster than exponentially with $1/r$ in agreement with the KAM theory. In the figures 9 we report the results for the tune shift for the non resonant linear frequency $\nu = .21$.

Fig.9

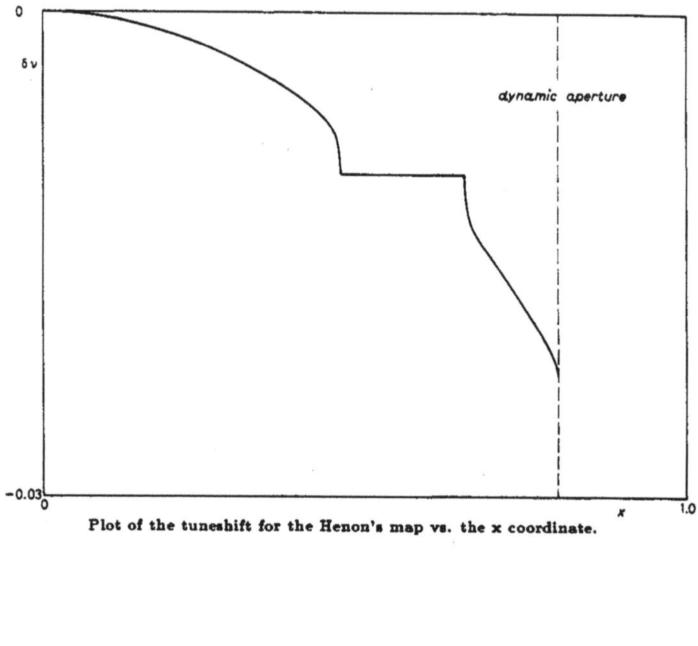

Plot of the tuneshift for the Henon's map vs. the x coordinate.

Of course in order to make a comparison we take the initial point $\mathbf{x} = (x, 0)$, we transform it into $\varsigma = \Phi(x, x)$ and then we compute $\Omega(\rho^2)$ where $\rho = |\varsigma|$. We recall that for the quadratic map in the nonresonant case the tune shift reads at the lowest order

$$\delta\nu = -\frac{1}{32\pi}\left[3\text{cotg}2\pi\frac{\nu}{2} + \text{cotg}2\pi\frac{3\nu}{2}\right]\rho^2 \tag{3.33}$$

The smear is defined as the deviation from circularity of the perturbed orbits. A possible definition is

$$\sigma_1 = \frac{r_{\text{ max}} - r_{\text{ min}}}{r_{\text{ max}} + r_{\text{ min}}} \qquad \begin{array}{l} r_{\text{ max}} = \max_{j\geq 1} \\ r_{\text{ min}} = \min_{j\geq 1} \end{array} \tag{3.34}$$

Another more convenient definition is given by

$$\sigma = \frac{\left\langle \left(r^2 - \langle r^2\rangle\right)^2 \right\rangle^{1/2}}{\langle r^2 \rangle} \tag{3.35}$$

On the orbit we have $r_j = |z_j| \equiv ||\mathbf{x}_j||$ and

$$\langle r^2 \rangle = \lim_{j\to\infty} \frac{1}{n}\sum_{j=1}^{n} r_j^2 \tag{3.36}$$

The normal form provides a parametric reprentation of the orbit; indeed letting $\varsigma = \rho e^{i\theta}$ we have

$$z = \Phi(\varsigma, \varsigma^*) = \sum_{n=1}^{N} \rho^n \sum_{l=0}^{n} a_{n,l} e^{i(2l-n)\theta} \tag{3.37}$$

where $a_{1,0} = 0$ and one can introduce the Fourier representation of $z = z(\theta)$ according to

$$z(\theta) = \sum_{-N}^{N} A_k e^{ik\theta} \qquad A_k = \sum_{n=|k|}^{N} \rho^n \, a_{n,\frac{k+n}{2}} \tag{3.38}$$

The Fourier spectrum itself of the orbit gives clear information on the deformation of the orbits. The smear is then calculated by simple algebraic relations involving the A_k. Indeed if we assume the tune to be non resonant, then the orbit is ergodic on the curve and we can write

$$\langle r^2 \rangle = \frac{1}{2\pi}\int_0^{2\pi} d\theta |z(\theta)|^2 = \sum_{k=-N}^{N} |A_k|^2 \tag{3.39}$$

and a similar expression for $\left\langle \left(r^2 - \langle r^2\rangle\right)^2 \right\rangle$. In figure 10 we show the smear for the non resonant frequency $\nu = .21$. For a resonant orbit the procedure is more involved since the orbits of the interpolating hamiltonian have to be considered.

4. Asymptotic properties of Birkhoff series and dynamic aperture

Fig.10

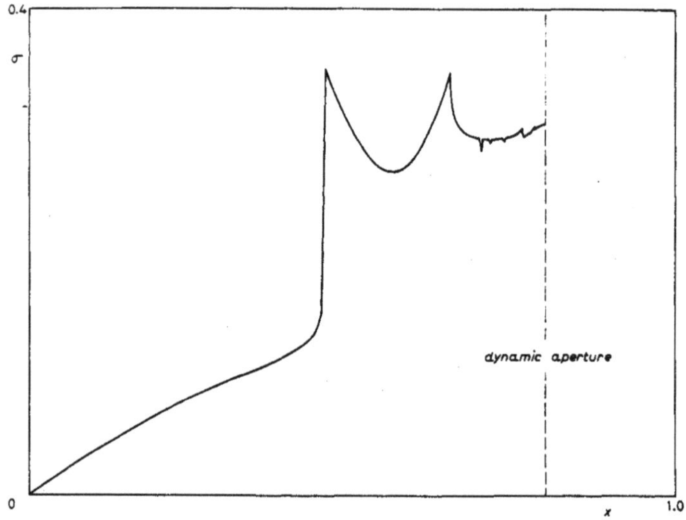

Plot of the smear for the Henon's map vs. x coordinate.

Let us transform the given map in normal form only at the lowest meaningful order ($N = 3$ if $\Omega_2 \neq 0$) and write it as

$$z' = e^{-i[\omega + \Omega_2 z z^*]} z + F_4(z, z^*) \tag{4.1}$$

where F_4 is a remainder of order 4. Letting $z = re^{i\theta}$ and p/q be a rational approximation to $\omega/2\pi$ where $2\pi\, p/q - \omega = \epsilon$, we see that for the integrable map any point of the circle of radius $r = \sqrt{\Omega_2/\epsilon}$ is a fixed point of order q. The full map will have a chain of q (or a multiple of q) islands. If for a given q we choose p as the closest integer to $q\omega/2\pi$ then $\epsilon \leq 1/2q$. However choosing $q = q_j$ one of the denominators of the continued fraction truncation of $\omega/2\pi$, we have $\epsilon \simeq 1/q_n^2$.

These *resonances* are the most relevant and govern the divergence of the Birkhoff series. Since a resonance of order q affects the series at order $N \geq q - 1$ we see that at order $q_j \leq N < q_{j+1}$ we should expect a convergent behaviour within the disc $r \sim 1/q_j$. Outside the failure of the normal form shows up in a dramatic way since the area preserving property of the conjugation function Φ is increasingly violated until Φ ceases to be invertible and the image of a circle is a curve with q_j loops, see fig 4. This is due to a fast increase of the Fourier coefficients A_k of the transformation Φ corresponding to the resonance namely for $k = \pm q_j + 1$ with respect to the non resonant coefficients.
We have already remarked that using the resonant normal form one can take into account only the first relevant resonance and its multiples so that the divergence problem remains.

The divergence mechanism of the series depends on the nonlinear nature of the functional equation and on the presence of the small divisors, which are related to the resonances. Indeed a leading divisor corresponding to the resonance p_j/q_j contributes according to

$$|e^{iq_j\omega}-1|^{-1}\sim[q_j\omega-2\pi p]^{-1}=[q_j\epsilon_j]^{-1}\sim q_j \tag{4.2}$$

The same contribution appears at any order m after the start up $N_j \geq q_j - 1$ and the norm of Φ_{N_j+m} behaves as q_j^m and appears to converge in a disk of radius $\rho \sim 1/q_j$ in agreement with our previous claim. Analytic estimates and numerical analysis confirm this behaviour. We observe that the resonance condition is given by equation (3.21) which has a real or imaginary solution for ρ. Of course the fixed points and islands show up only when ρ is real, nevertheless the apparent *convergence disc* is given by the modulus of the solution of (3.21), and is simply due to a cumulative effect of the corrisponding small divisor. It seems to us appropriate call the resonance *physical* in the first case and *virtual* in the second case.
If we take into account the cumulative effect of the resonances we see that asymptotically this piecewise geometric behaviour of the series implies a growth as N^{aN} for N large.

These heuristic arguments can made rigorous and recently a theorem was proved which shows that both the norms of Φ_N and Ω_N grow as $N^{a(N-2)}$ and that the remainder $\hat{R}_N(t)$ defined by

$$\Phi_N^{-1}\circ F^{ot}\circ\Phi_N-e^{it\Omega_N}\varsigma=\hat{R}_N(t) \tag{4.3}$$

is bounded by $t^2\,N^{a(N+1)}$, where $a=9+\eta$ and η is the diophantine exponent of ω namely

$$|e^{iq\omega}-1|^{-1}\leq\gamma q^{\eta} \tag{4.4}$$

If for a fixed r we look for the truncation order which makes the remainder as small as possible we find that $N_{\text{opt}}=e^{-1}(r_*/r)^{\frac{1}{9+\eta}}$ where $r_*=2/49\gamma$ provided that the norms of the coefficients of the map satisfy

$$||F_2||_1\leq 3^{-\eta}\gamma^{-1},\qquad ||F_n||_1\leq 1\ \text{for}\ n>2 \tag{4.5}$$

on the unit disc. The result holds for symplectic maps of $\mathbf{R}^{2d}$ and the optimal time, for which the orbit remains bounded can be estimated. The result is expressed by the following theorem proved in [16]

Theorem

If $F(z,z^*)$ is a symplectic map of $\mathbf{R}^{2d}$, with linear part $e^{i\omega}z$ where ω are non resonant frequencies satisfying a diophantine condition and such that (4.4) are verified, then assuming $||z^{(0)}||\leq r/2$ and letting $z^{(t)}=F(z^{(t-1)},z^{*(t-1)})$ one has $||z^{(t)}||\leq(1+\frac{1}{24})r$ for all $t\leq T$ such that

$$T=\frac{7}{8}\sqrt{\frac{r}{6}}\exp\left[\frac{9+\eta}{2e}\left(\frac{r_*}{r}\right)^{\frac{1}{9+\eta}}\right] \tag{4.6}$$

where

$$r_*=\frac{2}{49\gamma}\qquad r\leq r_*(3e)^{-9-\eta} \tag{4.7}$$

Even though the constants appearing in (4.6) are surely far from being optimal, the scaling law should be correct and numerical experiments are being carried out to check it. This is relevant in order to give a bound to the Arnold diffusion. The estimate provides an upper bound to the Arnold diffusion for maps of dimension at least 4 and a lower bound to the stability times. Rough numerical estimates of the diffusion show that it is very small and the stability times appear to be extremely long, not compatible with experimental results in beam experiments. As a consequence different models and mechanisms such as the modulation diffusion have to be considered.

Dynamic aperture

For the 2 dimensional maps no diffusion is present but the failure of the normal forms can also give very important information. The analysis is rather subtle: consider then order N non resonant normal and look for the boundary of the region $\hat{D}_\eta^{\,\mathrm{Jac}}$ where the transformation Φ has a significant deviation from simplecticity. To be more explicit this domain is given by:

$$\left|\frac{\partial(\Phi,\Phi^*)}{\partial(\varsigma,\varsigma^*)}-1.\right|\geq\eta \tag{4.8}$$

Such a domain will depend on N and ν as well. Let the $D_\eta^{\,\mathrm{Jac}}$ be the image by Φ of $\hat{D}_\eta^{\,\mathrm{Jac}}$. For a fixed N if ν is such that its closest dominant resonance is of order higher than N the region $\hat{D}_\eta^{\,\mathrm{jac}}$ is comparable with the basin of attraction of infinity. This is not surprising becouse in this case the only obstacle of topological nature to the existence of invariant curves are the hyperbolic points and the related hyperbolic tangle through which the points escape to infinity, namely the dynamic aperture itself. Of course when ν has a close dominant resonance of order less than N then of corse $\hat{D}_\eta^{\,\mathrm{Jac}}$ will touch the resonance region, which could be much inside the dynamic aperture. Consequenetly the result on the dynamic aperture would become more accurate as we increase N but the region in the tune axis where it can be applied becomes a multiintervall of measure decreasing with N. The dependence on η become weaker and weaker as N increases. Very similar results are obtained by considering the region $\hat{D}_\eta^{\,\mathrm{Rem}}$ where the remainder defined by (2.25) is above a given value $|R_N|\geq\eta$. The distance of the boundaries of $\hat{D}_\eta^{\,\mathrm{Rem}}$ from the origin is also comparable with the distance from the origin of the image of the set where Φ ceases to be invertible (typically a set of q points when ν has a dominant resonance of order q) specified by $d\theta/d\vartheta=0$ where $z=re^{i\theta}$ and $\xi=\rho e^{i\vartheta}$.

If we use the resonant normal forms the same considerations can be repeated; in this case for a given N the results the boundaries of the above sets will correspond to the dynamic aperture as far as the second dominant resonance is higher than N (see fig. 11).

In the case of $2d$-dimensional maps with $d>1$ the sets $\hat{D}_\eta^{\,\mathrm{Jac}}$ and $\hat{D}_\eta^{\,\mathrm{Rem}}$ can also be defined, however since in this case, due to the Arnold diffusion the basin boundary of ∞ is the origin, we have to consider them as regions of fast escape through hyperbolic manifolds without the slow down of narrow resonance channels.

Hyperbolic normal forms

If the tune is close to zero or 1/3 the dynamic aperture can be obtained, for the quadratic

Fig.11

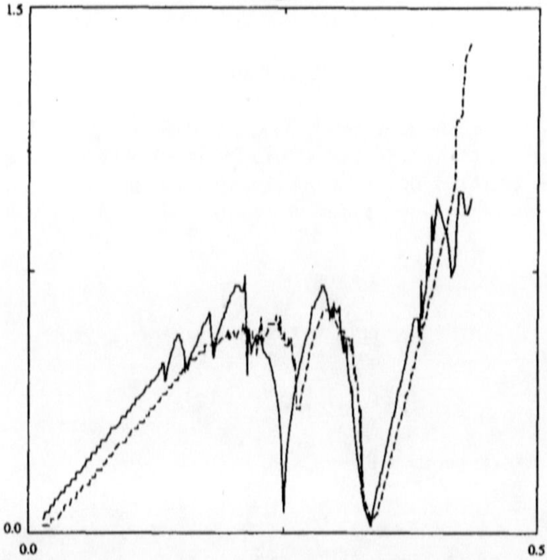

Dynamic aperture of the Henon's map vs. the linear frequency ν: coparison of the numerical computation (dashed line) with the Jacobian criterium of the normal forms (continuous line).

map by constructing inner envelope of the stable and unstable manifols emanating from the hyperbolic fixed points of F and $F^{\circ 3}$ and in principle for any dominant resonance of order q by the same construction for the fixed points of $F^{\circ q}$. The construction of these manifolds is allowed by the hyperbolic normal forms which converge according to the following theorem

Theorem (Moser 1956) Given an analytic area preserving mapping F with an hyperbolic fixed point at the origin then:

$$\exists \qquad \phi : \mathbb{R}^2 \to \mathbb{R}^2$$

s. t. the following diagram commutes

$$\begin{array}{ccccc} & (x,y) & \xrightarrow{F} & (x',y') & \\ \phi & \uparrow & & \uparrow & \phi \\ & (\xi,\eta) & \xrightarrow{U} & (\xi',\eta') & \end{array} \tag{4.9}$$

Where U is defined by:

$$U \quad \begin{cases} \xi' = \Lambda(\xi\eta)\xi \\ \eta' = \Lambda^{-1}(\xi\eta)\xi \end{cases} \tag{4.10}$$

and Λ, ϕ are convergent power series near the origin.

In the new coordinates we can construct explicitly the invariant, namely:

$$\xi'\eta' = \xi\eta$$

The convergence is due to the absence of the small divisors typical of the elliptic case. However this theorem is not sufficient to give us information about the motion. In fact, due to the unstable character of the fixed point after few iterations the point gets out of the convergence domain. Recently a result was proved which allows to overcome this difficulty [23]:

Theorem (Da Silva Ritter et al. 1987)

$$D^k = N^k(D_0), \qquad D^{-k} = N^{-k}(D_0)$$
$$\phi_k = F^{-k} \circ \phi \circ N^k \quad : D^{-k} \to \mathbb{R}^2 \tag{4.11}$$

where D_0 is the Moser convergence disk of Λ, ϕ, and

$$\mathbf{\Phi} = \lim_{n \to +\infty} \bigcup_{k=-n}^{n} \phi_k$$
$$\mathbf{D} = \lim_{n \to +\infty} \bigcup_{k=-n}^{n} D^k \tag{4.12}$$

Then $\mathbf{\Phi}$ brings F into normal form and is analytic on $\mathbf{D}$.

Because of the fact that $\mathbf{D}$ stretches along the ξ, η axes we can use this theorem to determine analytically the stable and unstable manifolds with their intersections. In the (ξ, η) plane these manifolds are simply the axes. In the original (x, y) plane, instead, they are no more straight lines, and are given by the preimages of the ξ, η through $\mathbf{\Phi}$. In order to show that the inner envelope of the hyperbolic manifolds gives the dynamic aperture we have considered the quadratic Henon map in the region $0 \le \nu \le .15$ dominated by the hyperbolic fixed point of F : $\mathbf{x}_0 = (\tan 2\pi\frac{\nu}{2}, -\tan^2 2\pi\frac{\nu}{2})$. In figure 12 we compare the dynamic aperture (boundary of the black region whose points are stable against iterations of the map) with the envelope of the hyperbolic manifolds and our assertion is fully confirmed.

The rigorous mathematical approach to the problem of dynamic aperture would be through the construction and characterization of the Julia sets of the fully complexified map, namely for the quadratic map

$$\begin{aligned} z' &= e^{i2\pi\nu}\left[z + \tfrac{i}{4}(z+w)^2\right] \\ w' &= e^{-i2\pi\nu}\left[w - \tfrac{i}{4}(z+w)^2\right] \end{aligned} \tag{4.13}$$

The Henon map is the restriction of (4.13) to the linear manifold $w = z^*$ and the dynamical aperture is the restriction of the Julia set of (4.13) to it. The consideration of the limit sets of families of complex trajectoris where the dynamics is the translation by

Fig.12

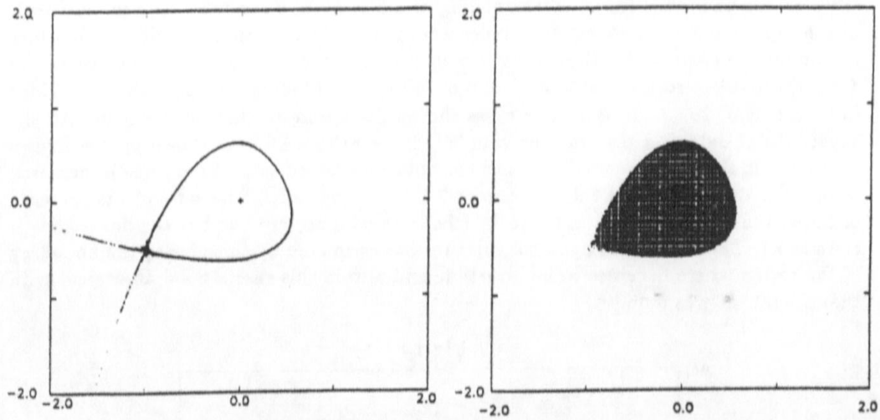

Hyperbolic invariant manifold(left) and dynamic aperture of the Henon's map.

$2\pi\nu$ seems also to provide interesting results related to the Julia set and are being investigated.

5 - Applications to hadronic lattices

The algorithms for the normal forms have been developed and numerically implemented first for the 2-dimensional maps [11], both in the non resonant and resonant case and later for the 4-dimensional maps in the non resonant case. A program for the maps of arbitrary dimensions is being developed.

The LHC tune shifs and smears

The first application considered has been the evaluation of the tune shift and smear for a LHC (Large Hadron Collider project at CERN) for the horizontal betatronic motion [18]. Later the same analysis has been repeated for the complete x, y betatronic motion and excellent agreement was found with the tracking results by FASTTRACK[19]. We observe that up to 20% of the dynamic aperture the computation of the tune shifts $\delta\nu_x, \delta\nu_y$ and smears σ_x, σ_y is much faster and accurate than by tracking. The perturbation order to be considered depends of the highest multipolar order taken into account: $2m$-pole contributes at order $N = m - 1$, and to the tune shifts at order $m - 2$ (becouse they are scaled by one order: $U = \varsigma e^{i2\pi(\nu+\delta\nu)}$).

The 1988 SPS experiment

The experiment carried out in the SPS in 1988, in order to evaluate the influence of the non linear fields on beam dynamic, was also analysed using the normal forms[21]. The beam was kicked horizontally and both tune shifts were measured at at some well defined amplitude. The short term dynamical amplitude, namely the amplitude at which no diffusion occurs for 10^5 turns, corresponding to a lifetime of 8 seconds was also measured. In the experiment eight strong sextupoles were powered with opposite polarities in order to minimize the effect of a third order resonance. The experiment provided three sets of data: the first two correspond to a current of 140 A. and different working points, the third to a current of 250 A. In the three cases the results are in excellent agreement with the experimental data and the tracking simulations by SIXTRACK as shown by the figures 13, 14, 15. In the first case no divergent behaviour is observed as the amplitude is increased whereas in the second and third the order 7 is no longer reliable becouse of the presence of the virtual resonance $(k_x, k_y) = (5, 0)$ (the tune does not cross it but the linear tune is close to it). The short term dynamic aperture was estimated by considering the boundary of the region where Φ ceases to be invertible and also in this case a good agreement with the experiment was found.

Fig.13

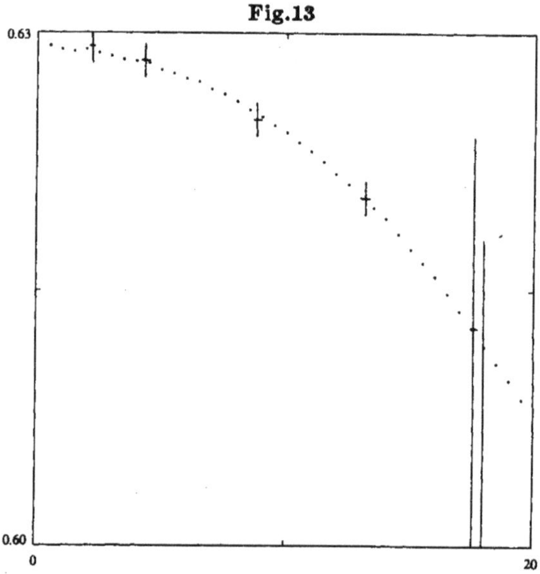

Horizontal tune with normal form up to order 7 in experiment n. 1; the crosses are the experimental data, the bares denote the short term dynamic aperture: the left one is an experimental value, the right one is an analytical estimate with the normal form.

The correction of errors in LHC

The correction of the multipolar errors is a crucial point in the design of the new large hadronic machines such as LHC, SSC, ELOISATRON. The normal form approach is certainly the most promising and effective. The correction strategy is not a priory obvious. Indeed the ideal procedure would consist in defining the correctors in order to maximaze

Fig.14

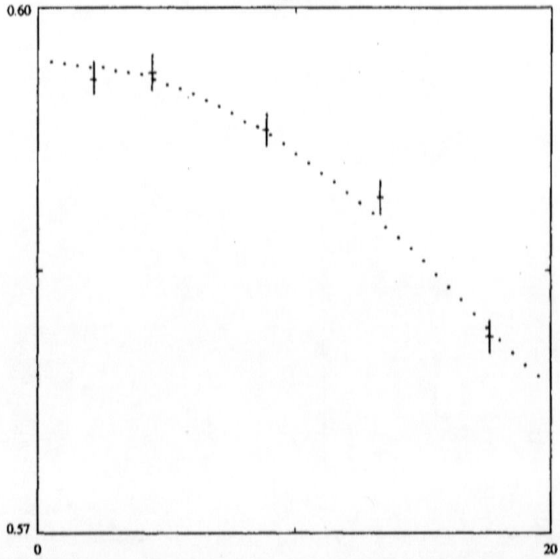

Horizontal tune with normal form up to order 5 in experiment n. 2; the crosses are the experimental data.

the short term dynamical aperture. This is the most difficult quantity to be computed by analytical tools; as a consequence since the computation of the tune shift and smear is accurate in the inner region one can try to minimize one of them. The tune shift is to be preferred becouse it is an invariant along the lattice while the smear varies with the chosen reference plane (Poincaré section). The smear has already a more complicated expression with respect to the tune shift and a simultaneous computation on all the lattice elements (σ_j in order to carry out a simultaneous minimization (namely of $\sigma = \sum_j \sigma_j$) would make it remarkably more time consuming.
As a first example we consider a cubic map where the quadratic term is fixed and the cubic one is varied and can be thought as the corrector.

$$z' = e^{-i2\pi\nu}\left[z - \frac{i}{4}(z+z^*)^2 + i\alpha(z+z^*)^3\right] \tag{5.1}$$

The second order expression of the tune shift is

$$\delta\nu = -\frac{1}{32\pi}\left[3\cot(\pi\nu) + \cot(3\pi\nu)\right] + \frac{3}{2\pi}\alpha \tag{5.2}$$

At this order the tune shift can be made equal to zero and one has

$$\alpha = \frac{1}{48}\left[3\cot(\pi\nu) + \cot(3\pi\nu\right] \tag{5.3}$$

Fig.15

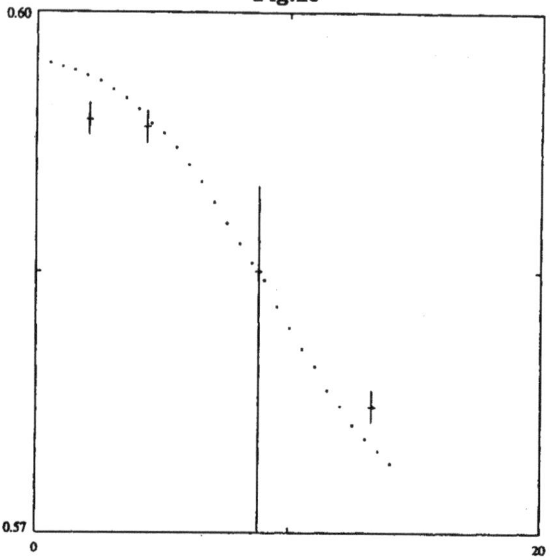

Horizontal tune with normal form up to order 5 in experiment n. 3; the crosses are the experimental data and the bare denotes the experimental value of the short term dynamic aperture.

In a real lattice the tunes are obtained at second order by differentiating with repect to $\rho_1 = \varsigma_1\varsigma_1^*$ and $\rho_2 = \varsigma_2\varsigma_2^*$ a function $\mathcal{H} = \nu_1\rho_1 + \nu_2\rho_2 + h_{20}^2 + h_{21}\rho_1\rho_2 + h_{22}\rho_2^2 + \ldots$ and the tune shifts read

$$\begin{aligned}\delta\nu_x &= 2h_{22}\rho_1 + h_{21}\rho_2 + \ldots \\ \delta\nu_y &= h_{11}\rho_1 + 2h_{20}\rho_2 + \ldots\end{aligned} \tag{5.5}$$

The dependence of the multipolar coefficients in second and fourth order (in $|\varsigma|$ or first in ρ) is the following

$$\text{second order } h_{2,i} \propto \begin{cases} k_2^2 & \text{sextupole field} \\ k_3 & \text{octupole field}\end{cases}$$

$$\text{fourth order } h_{3,i} \propto \begin{cases} k_5 & \text{dodecapole field} \\ k_2k_4 & \text{decapole+sextupole field} \\ k_3k_2^2 & \text{octupole+ sextupole field} \\ k_3^2 & \text{octupole field} \\ k_2^4 & \text{sextupole field}\end{cases} \tag{5.6}$$

Since the $h_{n,i}$ are multilinear in the integrated gradients k_j, the corresponding coefficients are obtained by solving a system of linear equations obtained by computing the $h_{n,i}$ and assigning suitable values 0 or 1 to the k_i.

As an example we consider a FODO cell, see figure 16, where we include for each multipolar error a corrector after the focusing quadrupole a central corrector and a corrector after the defocusing quadrupole.

Fig.16

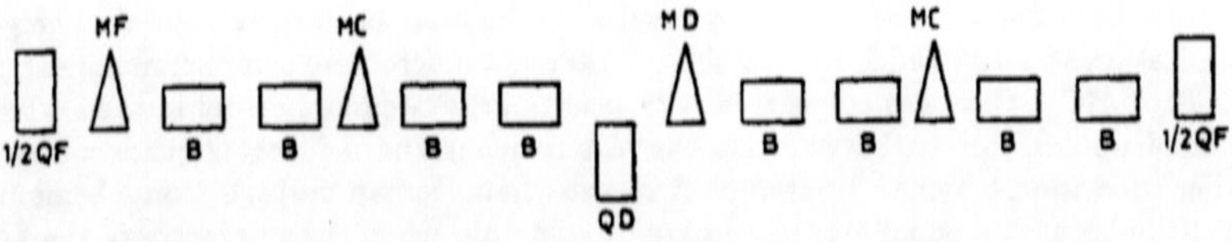

Scheme of a FODO cell with correctors: the B indicates a dipole magnet, the Q a quadrupole and the M a multipole.

To the second order tune contribute only the sextupoles and octupoles namely

$$k_{2F}, k_{2C}, k_{2D}, k_{3F}, k_{3C}, k_{3D}$$

where k_{2F} and k_{2D} are fixed by imposing that the cromaticity is corrected. As a consequence we are left with 4 parameters and we write

$$h_{2,i} = \alpha_i + \beta_i k_{2C} + \gamma_i k_{2C}^2 + \delta_i k_{3C} + \eta_i k_{3F} + \epsilon_i k_{3D} + \dots \qquad i = 0, 1, 2 \tag{5.7}$$

At fourth order corresponding to the 5th order of the map we have also

$$k_{4F}, k_{4C}, k_{4D}, k_{5F}, k_{5C}, k_{5D}$$

namely 6 additional parameters.
The correction strategy is simple. We minimize the tune shift in the polydisc $0 \leq \rho_1 \leq R, 0 \leq \rho_2 \leq R$ in two different ways: 1) order by order (namely we minimize the function)

$$\int_0^R d\rho_1 \int_0^R d\rho_2 \, [\delta\nu_x|_{2\ell}]^2 + [\delta\nu_y)|_{2\ell}]^2 = R_{2\ell}\chi_{2\ell}(k_j)) \qquad \ell = 1, \dots, m \tag{5.8}$$

2) globally

$$\int_0^R d\rho_1 \int_0^R d\rho_2 \sum_{\ell=1}^{m} ((\delta\nu_x|_{2\ell})^2 + (\delta\nu_y)|_{2\ell})^2) = \sum_{\ell=1}^{m} R_{2\ell}\chi_{2\ell}(k_j) \tag{5.9}$$

We observe that $h_{2,i}$ is linear in k_3 and quadratic in k_2, $h_{4,i}$ is linear both in k_3 and k_4 and in general $h_{2m,i}$ is linear in $k_{2m-1,i}$ and $k_{2m,i}$. As a consequence the order by order minization implied the solution of a non linear problem only when the second order is considered. Indeed χ_2 will have a quadratic dependence on k_{3F}, k_{3C}, k_{3D} and up to quartic in k_{2C}. The minimization ends up to the solution of 3 linear equations for the octupolar correctors and a cubic equations for the sextupolar corrector. If we fix a symmetry condition such as $k_{3D} = k_{3D}$ it possible obviusly to make the tune shift exactly zero. At second order we have only to solve linear minimization equations for the three decapolar and the three dodecapolar correctors.

The order by order correction strategy results can be used as starting point for the global minimization strategy which implies the solution of a much more complex nonlinear problem. The LHC lattice was considered with insertions, systematic errors in the sextupoles and quadrupoles, chromaticity corrections with focusing and defocusing quadrupoles. The Neuffer correction scheme[24] based on first order hamiltonian perturbation scheme where all multipoles are on equal footing, proves to hold only when the symmetry of the FODO cell is not broken by insertions and only octupolar errors are corrected with octupoles. Three correction schemes are proposes[25]

a) $k_{iF} = k_{iD}$ $\quad i = 2,3,4$: 5 parameters
b) $k_{iF} = k_{iD}$ $\quad i = 2,3,5$: 7 parameters
c) $k_{iF} \neq k_{iD}$ $\quad i = 2,3,4$: 7 parameters

which allow to widen the zone in the phase space where the motion has a linear behaviour. We conclude that the method is automated to work up to very high order and it is extremely fast and reliable, the only care to be taken being the behaviour of the tune shift near the dynamic aperture. We report in table 1 the ratio between the tuneshift of the bare machine and of the corrected machine at different amplitude for the different correction schemes.

Table 1

Amplitude (mm)	Scheme a	Scheme b	Scheme c
2	60	700	350
4	40	150	45
7	4.5	3.5	4

REFERENCES

[1] E. Courant, H. Snyder Annals of Physics (N.Y.) **3**,1 (1958)
[2] W. Scandale CERN-SPS report 87-28 (AMS) (1987)
[3] F. Schmidt Ph.D thesis DESY HERA report 88-02 (1988)
[4] R. D. Routh "Single particles dynamics and nonlinear resonances in circular accelerators" from *Non linear dynamics aspects of particle accelerators* Ed. by Jowett,Month and Turner, Springer-Verlag pag. 37 (1985)
[5] A. Bazzani, A. Pisent, G. Turchetti *Normal forms for hamiltonian maps and future applications to accelerators* CERN report SPS 87/38(AMS) (1987)
[6] A. Dragt, J. M. Finn J. Math. Phys **17** pag. 2215 (1976)
[7] G. D. Birkhoff Acta Mathematica bf 43, 1 (1927)
[8] V. I. Arnold, A. Avez *Ergodic Problems of Classical Mechanics* Ed. Benjamin (1968)
[9] J. Roels, M. Henon Bull. Astr. Soc. **2**, 267 (1967)
[10] G. Benettin, A. Giorgilli, G. Servizi, G. Turchetti Phys. Letters **A95**, 11 (1983).
[11] G. Servizi, G. Turchetti Il Nuovo Cimento **B95**, 121 (1986)
[12] A. Bazzani, G. Turchetti "Normal forms for symplectic map" from *Nonlinear Dynamics and stochastic processes II* Ed. by Livi, World Scientific pag. 25 (1988).
[13] A. Bazzani Celestial Mechanics **42**, 107 (1987)
[14] G. Turchetti "Perturbative methods for hamiltonian maps", from *Methods and applications of nonlinear dynamics* Ed. W. A. Saenz. World Scientific Publ. pag. 95 (1988).
[15] G. Turchetti "Nekhoroshev stability estimates for symplectic maps and physical applications" from *Number theory in Physics* ed. P. Moussa, Springer (1989)
[16] A Bazzani, S. Marmi, G. Turchetti "Nekhoroshev estimates for non resonant symplectic maps" Celestial Mechanics, to be published (1989)
[17] M. Pusterla, G. Servizi, G. Turchetti, IEEE Transactions on nuclear sciences **NS30**, 2642 (1983).
[18] A. Bazzani, G. Turchetti, P. Mazzanti, G. Servizi Il Nuovo Cimento **B102**,51 (1988)
[19] A. Bazzani, B. Leemann, W. Scandale, G. Servizi, G. Turchetti "Description of nonlinear beam dynamics in the CERN large hadron collider by using normal form algorithms" from *European Particle Accelerators Conference (EPAC 1988)* Ed. S. Tazzari, World scientific (1989)
[20] E. Forrest, M. Berz, LBL Accelerator & Fusion Research Division report 25609 UC-414 (1988)
[21] A. Bazzani, G. Turchetti "Normal form results for the 1988 dynamical aperture experiment" CERN SPS report 89-25 (AMS) (1989)
[22] W. Scandale, F. Schmidt, E. Todesco CERN SPS report (1989)
[23] G. L. Da Silva Ritter et Al. Physica D **29** pag. 181 (1987)
[24] D. Neuffer "Lumped multipole correction for the CERN Large Hadron Collider" CERN report 87-014 (AHF) (1987)
[25] E. Todesco "Fome normali risonanti per mappe simplettiche polinomiali e correzioni multipolari per LHC" Tesi di Laurea dell' universitá degli studi di Bologna (1989)

LOW-DIMENSIONAL EVOLUTION OF NONLINEAR DYNAMICAL SYSTEMS

Antonio Politi

Istituto Nazionale di Ottica, I-50125 Firenze, Italy

Abstract. The general techniques introduced in the last years to characterize the low-dimensional dynamical behavior of nonlinear systems, are reviewed. The concepts of fractal dimension, Lyapunov exponents and metric entropy are analyzed, discussing also the relations intercurring among them. Finally, the state of arts for hyperbolic and non-hyperbolic systems is briefly summarized.

1. Introduction

Since the first numerical evidence of deterministic chaos, presented by Lorentz in his famous paper [1], a large number of (both numerical and experimental) observations of low dimensional chaos has been reported in many different areas of scientific research: They range from optics, to biology, chemistry, acoustics, hydrodynamics, to speak of few examples. It is now widely accepted that the observation of an erratic behavior is not to be necessarily associated with the coupling of a given system with a thermal bath (schematized as an infinite number of external uncontrollable degrees of freedom). Rather, it often follows from the existence of a nonlinear coupling among a few variables. In such cases, the number of nonnegligible Fourier modes in the frequency spectrum cannot be any longer considered as a measure of the "complexity" of a given temporal signal. In fact, it is in principle possible to construct a set of a few ordinary differential equations which reproduce the observed signal. As a consequence, the number of effective degrees of freedom engaged in the asymptotic dynamics seems to be a more appropriate indicator of dynamical "complexity". Such a number roughly corresponds to the fractal dimension of the underlying strange attractor, an indicator which can be extracted from experimental data.

The problem of characterizing deterministic chaos can be attacked along different lines: (a) by describing the invariant measure in phase space in terms of an ensemble of fractal dimensions; (b) by determining the divergence rate of nearby orbits in terms of the Lyapunov (characteristic) exponents; (c) by following an information theoretical approach, computing metric and topological entropy as a measure of the production rate of information.

However, prior to discuss deterministic chaos, it is useful to spend a few words on the problem of understanding why, in many circumstances, a few variables do provide a sufficiently accurate description of the asymptotic behavior of a dynamical system. Indeed, very often, a starting realistic model is given in terms of a set of partial differential equations, which imply an infinite-dimensional phase space. This is the case of hydrodynamics where the Navier-Stokes equations (possibly coupled with heat diffusion equation) represent the most suited model to describe e.g. the turbulent motion in a channel, or the Rayleigh-Benard instability. Quantum optics represents another field of physics where a rather accurate description of the temporal evolution of a laser source is provided by the Maxwell-Bloch equations which describe, in a self-constient manner, the dynamical behavior of an electric field interacting with a set of polarizable atoms. Reaction diffusion equations, describing various chemical instabilities like Beluzov-Zabotinsky reaction, belong again to the class of partial differential equations. Despite the highly complex structure of the starting models, their evolution can be very trivial. A fluid heated from below exhibits, under the convection threshold, a stationary state with temperature and velocity fields depending only on the spatial variables. In a standard laser just above threshold, the output intensity is constant, exactly as the asymptotic concentration of chemical compounds in many reactions. The explantation of such behaviors is intrinsically connected with the dissipative character of the above mentioned systems: an initially finite volume in phase space shrinks to zero with an exponential rate given by the divergence of the equations. A stationary state is therefore the most likely state to be asymptotically reached. Indeed, for a long time it was believed that we should have expected at most periodic oscillations, or a combination of oscillations with different frequencies (this was indeed the idea of Landau on the onset of turbulence). Only more recently, long after the numerical experiments of Lorentz [1], physicists started realizing that the asymptotic attractor could be other than a closed curve in the phase space.

Before entering the problem of describing these "strange" sets, I want to discuss the reason why a few degrees of freedom can, anyhow, be sufficient to provide an accurate description of a given system. I start noticing that in many instances not only a global dissipation occurs in phase space, but, more specifically, many variables undergo a fast decay towards an equilibrium position. This is for instance the case of the atomic polarization which, in many lasers, rapidly converges to a local equilibrium position as if the electric field amplitude and population inversion were constant in time. This is because the characteristic decay time of such quantities is much longer than the relaxation time of polarization. In this and similar cases the "fast" variables are adiabatically eliminated, as they do not contribute to the asymptotic evolution. By using Haken's terminology, the fast variables are slaved by the slow ones [2]. As a consequence, only the (few) slow degrees of freedom are responsible for the evolution of a typical physical system. However, the reader must be warned against a blind application of adiabatic elimination. Historically, the method has been introduced with reference to weakly interacting systems, each characterized by a proper decay time. In such a case, once the various relaxation times have been mutually compared, it is sufficient to set equal to zero the time derivative of all the variables characterized by a fast decay process. This is for instance the approach which allows us to pass from the Maxwell-Bloch to the rate equations. However, when the coupling becomes sufficiently strong, the picture based on the decay times of the original variables can prove to be rather ineffective. This can be already clarified by considering linear systems. Weak interactions means that the force field is given by a quasi-diagonal matrix, so that the decay rates (diagonal terms) yield a good approximation of the eigenvalues of the matrix. It is obvious that strong interactions leading to generic matrices, can strongly affect the eigenvalues. However, as long as a linear approch is approximately correct, a suitable change of variables, which diagonalizes the force field, solves the problem, allowing us to determine the truly fast variables to be eliminated which thus turn out to be combinations of the original ones. This is just the first step of centre manifold theory [3], which is the best approach so far developed to reduce the number of degrees of freedom in a quantitatively controllable way.

The starting point of such a theory is the observation that the existence of fast variables, each one characterized by a dissipative evolution, implies the existence of an invariant (and in general nonlinear) manifold in phase space, where the asymptotic motion occurs. The exact determination of such a (centre) manifold allows us to reduce the dimensionality of the model, as the number of relevant degrees of freedom is now given by the dimension of the centre manifold. What is lost by the resulting model is only the information on the convergence to the asymptotic dynamics starting from a generic initial condition outside the manifold. We can think of this theory as the extension of diagonalization technique to nonlinear dynamical systems. The reason why we have to limit ourselves to factorize the evolution between slow and fast variables, rather than considering separately all degrees of freedom, is twofold.

First, it is in general possible to determine the centre manifold only in a perturbative way, so that only when the time scales of the relaxation processes are well separated, there is a fast convergence. Unfortunately, we have to recall that, in many cases, the time scales of the relaxation phenomena cover a whole range in a continuous fashion so that a quantitative analysis can be hardly worked out and only phenomenological analyses can be attempted.

Second and more important, the existence of chaos prevents the possibility of decoupling the motion in phase space into independent evolutions of different degrees of freedom. This latter observation means that we have to stop the reduction procedure to the point when the asympotic motion "fills" the manifold.

2. Deterministic chaos

Assuming that all irrelevant relaxation phenomena have been eliminated from the model, we want to ask ourselves which kind of motion can be expected to be found. We assume that all stationary (fixed points) and periodic solutions are unstable, and

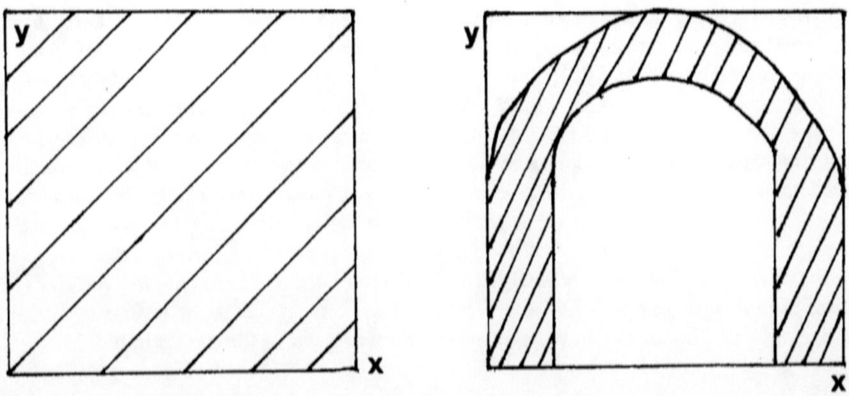

Fig. 1: Smale's horseshoe - a two-dimensional map exhibiting the two basic ingredients to generate chaotic behavior: stretching along an (unstable direction) and folding to guarantee confinement in a finite region of phase space

ask ourselves whether it is possible to construct a model exhibiting a motion confined in phase space. A linear stability analysis can only tell us that the state of the system escapes from the unstable solutions. A global qualitative analysis is more suited to attack the problem. The fundamental role played by nonlinearities is immediately clear, by noting that they provide the necessary ingredient to confine the phase point inside a finite region of phase space, nothwithstanding the instability of any periodic solution: *folding*. In fact, since we assume to investigate a dissipative system, a finite region in phase space, stretched along a suitable (unstable) direction, is also squeezed along a (stable) direction. Accordingly, only a folding mechanism can prevent to solution to explode to infinite. These are the basic ingredients contained in the horsehoe map proposed by Smale [4] and sketched in Fig. 1 (from now on, the time will be assumed to be discrete without loss of generality). It guarantees simultaneously a sensitivity to initial conditions (one positive Lyapunov exponent) and the confinement of the orbit. The inaccuracy due to a non perfect knowledge of the initial conditions is transformed into a seemingly erratic evolution in phase space, which in turn naturally suggests the statistical approach as the most appropriate one.

However, before entering the problem of developing a suitable formalism, it is important to comment on the real occurrence of chaotic behavior in systems characterized by a stretch-and-fold dynamics. If it were true that, everywhere in the portion of interest of phase space, the expanding direction is transversal to the contracting one (assumption which defines the class of the so called *hyperbolic systems*), then it would be easy to convince oneself that a structurally stable erratic behavior is to be expexcted. A generic (and sufficiently small) perturbation of the dynamical laws cannot modify the transversality property of stable versus unstable manifold. As a consequence, such a nice property, allowing us to control the degree of chaoticity, makes also possible to prove rigorously many relationships among the various indicators of chaos [5]. In fact, practically everything which is nowadays rigorously known for dissipative chaotic systems, is restricted to hyperbolic systems. Unfortunately (or fortunately), no one realistic model is hyperbolic. The continuity of dynamical laws impedes the transversality of the manifolds. This can be already understood with reference to the Smale horseshoe where it is heuristically seen that the (almost everywhere vertical) expanding direction is, in some region, mapped into the horizontal contracting one, and viceversa. Therefore, it is not at all a priori obvious if an instability can asymptotically develop, after infinitely many of such exchanges occurred. How the situation is intrigued is testified by the few analytical results which have been proved for the class of non-hyperbolic systems. The simplest sub-class is represented by unimodal maps of an interval onto itself, the best known example of which is the "logistic map" ·

$$x_{n+1} = 1 - ax_n^2 \quad . \tag{2.1}$$

For such transformations, windows of finite length in parameter space where a stable periodic solution exists can be found arbitrarily close to any parameter value. Nevertheless it has been proved that chaotic orbits exist on sets of parameter values of positive Lebsegue measure [6]. The problem was much more disputed already for the Hénon map

$$x_{n+1} = 1 - ax_n^2 + bx_{n-1} \quad , \tag{2.2}$$

which is very similar to the Smale horseshoe, and reduces to the logistic map for $b = 0$. The source of difficulties is the possible occurrence of tangencies between the stable and unstable manifolds (homoclinic tangencies). The effect of the tangency is very similar to the vanishing slope in the logistic map and leads to the Newhouse phenomenon [7]: even if a chaotic attractor exists at a parameter pair (a, b), there is an infinity of attracting periodic orbits arbitrarily close to it. Only very recently, it has been proved that in a suitable region of parameter space (not including the canonical values $a = 1.4$, $b = 0.3$), chaotic behavior is expected to occurr on a set of points of finite Lebesgue measure [8].

3. Dimensions

When dealing with fractal objects on which a measure μ is defined, the dimension D is introduced to describe the increase of mass $p(\varepsilon)$ with size ε [9]

$$p(\varepsilon) \simeq \varepsilon^D \tag{3.1}$$

In general, however, fractals are not uniform, i.e. the mass scales differently from point to point. Therefore one is lead to consider pointwise dimensions

$$\alpha(x) = \lim_{\varepsilon \to 0} \frac{\log p(\varepsilon, x)}{\log \varepsilon} \tag{3.2}$$

where $p(\varepsilon, x)$ is the mass of an ε-ball centered at x. In order to give a global estimate of dimension, it is then necessary to define a suitable average of the "local" dimensions

$\alpha(x)$. By referring to a covering of the fractal object with balls of variable size ε_i, centered at points chosen at random with respect to the measure μ (rather than to the Lebesgue measure, to avoid the difficulties associated with boxes containing spuriously small masses), the generalized (Renyi) dimensions $D(q)$ are defined by [10]

$$\left\langle \left(\frac{p}{\varepsilon^{D(q)}}\right)^{(q-1)} \right\rangle \simeq \text{Const.} \quad , \quad \text{for all } \varepsilon_i \to 0 \tag{3.3}$$

For $q = 0$ we obtain a purely geometrical quantity (box dimension), while in the limit $q \to 1$ the so called information dimension is obtained. Eq. (3.3) can be simplified in essentially two complementary ways. In fact, we can use balls of fixed radius, deriving

$$\langle p(x)^{q-1} \rangle \simeq \varepsilon^{\tau}(q) \quad , \quad \text{for } \varepsilon \to 0 \tag{3.4}$$

where

$$\tau(q) = (q-1)D(q) \tag{3.5}$$

This is the base of the Grassberger-Procaccia algorithm to compute the correlation exponent [11]. Alternatively, one can take balls of equal weight p and variable radii ε_i. This leads to

$$\langle \varepsilon(x)^{-\tau(q)} \rangle \simeq p^{(1-q)} \quad , \quad \text{for } p \to 0 \tag{3.6}$$

which is the base of the algorithm for extracting the fractal dimensions from the distance of the nearest neighbours [12].

The characterization of nonuniformity of the previous subsections can be reformulated by introducing a scaling function $f(\alpha)$ describing the spread of values assumed by the pointwise dimensions (3.2) [13]. Introducing the probability density $P(\alpha;\varepsilon)$ of pointwise dimensions (we still call α a pointwise dimension although the limit $\varepsilon \to 0$ clearly has not yet been taken)

$$P(\alpha;\varepsilon)d\alpha = Prob\{\alpha(\varepsilon, x) \in [\alpha, \alpha + d\alpha]\} \tag{3.7}$$

we can rewrite Eq. (3.4) as

$$\int d\alpha P(\alpha;\varepsilon)\varepsilon^{(q-1)\alpha} \simeq \varepsilon^{\tau(q)} \quad , \quad \text{for } \varepsilon \to 0 \tag{3.8}$$

This scaling law is satisfied by the Ansatz

$$P(\alpha;\varepsilon) \simeq \varepsilon^{\alpha - f(\alpha)} \tag{3.9}$$

where the balls have been chosen according to their own probability. In the limit $\varepsilon \to 0$, the integral is dominated by the value $\alpha(q)$ of α that yields the minimum of the expression $q\alpha - f(\alpha)$. Therefore, one has

$$\tau(q) = \min_{\alpha}\{q\alpha - f(\alpha)\} \quad , \quad f(\alpha) = \min_{q}\{q\alpha - \tau(q)\} \tag{3.10}$$

i.e. the function $f(\alpha)$ is the Legendre transform of $\tau(q)$. In particular, if $f(\alpha)$ and $\tau(q)$ are differentiable, one obtains

$$f(\alpha) = q\alpha(q) - \tau(q) \quad , \quad \alpha(q) = \frac{d\tau(q)}{dq} \quad , \quad q = \frac{df(\alpha)}{d\alpha} \tag{3.11}$$

The maximum of $f(\alpha)$ is the box dimension $D(0)$, while $f(D(1)) = D(1)$ and $f'(D(1)) = 1$. Instead of applying Eq. (3.4) we could have used relation (3.3), with p as a scaling parameter. For a more complete review on the subject see Ref. [14].

The above sketched theory of dimensions allows to reconstruct all the properties of a fractal measure only in the case of sets embedded in a one-dimensional space or for an isotropic distribution. In fact, in both cases, a scalar quantity like the radius ε of a ball is sufficient to identify the only relevant scaling index α (from Eq. (3.2)). In general, we expect to find a more intrigued situation. In the case of a Cartesian product of two Cantor-like sets, each point of the fractal set is fully described, by construction, by two indices α_1, α_2 accounting for the Cantor structure along two suitable transversal directions. Therefore, for this class of systems, we can still use the same kind of formalism with some modifications. Instead of spheres we consider ellipsoids with semi-axes of variable length. This is tantamount to considering ε as a vector-like quantity, as well as the pointwise dimension α and the Renyi dimensions. In other words it is useful to introduce the partial dimensions whose sum yields the global quantities defined in the first part of this Section. The success of this strategy depends simply on the fact that an anisotropic object is reduced to the "product" of many independent isotropic objects. We do not expect this to be true for generic fractal measures we can encounter in physics. In such cases a truly multi-dimensional technique is required to take into account the correlations among various scaling indices.

As long as strange attractors are involved, it has been conjectured that their local structure is nothing but the Cartesian product of various continua and one Cantor-like set. Accordingly, the above mentioned generalization of the usual approach should be sufficient to describe their invariant measure. This conjecture appears to be definitely correct for what concern the unstable manifold in hyperbolic systems, where the expansion process leads to a continuous distribution. However, it is not clear whether it still holds for the stable manifold whenever it contributes with a value larger than 0 to the global dimension. In such a case, as shown in Ref. [15], it is certainly true that the attractor is not the Cartesian product of some continua by a Cantor-like set, so that it is not clear whether we can still speak of partial dimensions in some weaker sense. Finally, for non-hyperbolic attractors, it is not true that the invariant distribution along the unstable manifold is uniform. In the logistic map there are square-root singularities all over the interval which decrease the value of $D(q)$ below 1 for $q > 2$, thus inducing a discontinuity in its derivative. An analogous behavior of $D(q)$ has been numerically observed and euristically explained in two-dimensional maps like the Hénon map for q larger than a critical value q_c [16]. As a consequence we can conjecture that non-hyperbolic systems are characterized by two distinct phases: (a) a hyperbolic one, where the known results apply (e.g. the Cartesian product structure); (b) a non-hyperbolic one made of homoclinic tangencies, where an alternative formalism must be developed, and where the concept of partial dimensions seems to be inapplicable.

This rough idea of distinct phases can be substantiated through the interpretation of the fractal dimension formalism in terms of a more familiar statistical mechanics formalism [17]. For sake of clarity, we consider a non trivial Cantor-like fractal (e.g. the Feigenbaum attractor). In this case, it is convenient to cover the fractal set with boxes of constant mass (different procedures lead to different identification schemes with thermodynamic quantities [18]) $p = 2^{-h}$. This is immediately done by considering the set at the h-th hierarchical step of its generation procedure, where it is composed of 2^h bands of variable width, which are exactly matched by the covering boxes. Therefore, we note that each box can be identified by a sequence $\{S_h\}$ of h 0's and 1's; that is, the sum in Eq. (3.6) can be interpreted as the sum over all possible configurations of a one-dimensional lattice gas of h particles. More precisely, we can make the following sequence of identifications

$$\begin{aligned} V &= -\log p \\ \beta &= \tau(q) \\ F &= 1/D(q) \\ E &= 1/\alpha \end{aligned} \tag{3.12}$$

where V represents the volume, $\beta \equiv 1/k_B T$ with T being the temperature and k_B the Boltzmann constant, F is the free energy and E is the energy of each single configuration. According to Eq. (3.12), relation (3.6) can be interpreted as the well known definition of the canonical partition function

$$Z_{can} \equiv \sum_{\{S_h\}} \exp[-\beta E(\{S_h\})] = \exp[-\beta F] \tag{3.13}$$

By proceeding further on, $f(\alpha)$ is easily identified with the entropy. Now, our previous vague reference to distinct phases can be more rigorously understood. The occurrence of a discontinuity of the derivative of $D(q)$ can be interpreted as a phase transition in the thermodynamic language.

4. Lyapunov exponents

In the previous Section we have considered quantities characterizing scaling properties of generic fractal sets. When studying strange attractors, one has the additional information coming from the underlying dynamics, and, in particular, Lyapunov exponents can be used to estimate dimensions and metric entropies.

In order to study these connections, it is useful to introduce the effective Lyapunov exponents, by referring to large but finite times [19]. More precisely, we consider an infinitesimal ball of radius $\varepsilon(x, n = 0)$, centered at x at time 0. Some time n later it will be transformed into an ellipsoid with semiaxes $\varepsilon_j(n)$, $j = 1, \ldots, E$ (E being the dimension of phase space). Effective Lyapunov exponents are defined as

$$\lambda_j(n, x) = \frac{1}{n} \lim_{\varepsilon \to 0} \log \frac{\varepsilon_j(n)}{\varepsilon(x, 0)} \tag{4.1}$$

Now, analogously to the pointwise dimensions, we discuss the probability distribution $P(\lambda, n)$ of effective Lyapunov exponents in the limit $n \to \infty$, where we have introduced the vector notation $\lambda = (\lambda_1, \ldots, \lambda_E)$. This can be done by following two equivalent routes: (a) via the cumulant generating function, which leads to the generalized Lyapunov exponents (analogously to the Renyi dimensions); (b) directly, in terms of an Ansatz on the scaling behavior of $P(\lambda, n)$, which leads to the distribution of effective Lyapunov exponents, to be confronted with $f(\alpha)$.

For sake of brevity, here we limit ourselves to summarize the first approach. From the moments of the multipliers $e^{\lambda n}$, we can introduce

$$g(\mathbf{z}) = \lim_{n \to \infty} \frac{1}{n} \log \int d\mu(x) \exp[n\mathbf{z} \cdot \lambda(n, x)] \tag{4.2}$$

where $\mathbf{z}$ is a vector the components of which play the role of $q - 1$ in the theory of fractal dimensions, and $g(\mathbf{z})$ is to be confronted with $\tau(\mathbf{q})$. However, at variance with $\tau(\mathbf{q})$ which is, by definition, the scalar product of the vector $(\mathbf{q} - 1)$ with the generalized dimension vector $\mathbf{D}(\mathbf{q})$, here we cannot factorize $g(\mathbf{z})$ in a similar way, because of the correlations existing among the various directions. A consistent definition of generalized Lyapunov exponents is

$$\Lambda_j(z_j) = \frac{g(0, 0, \ldots, z_j, \ldots, 0)}{z_j} \tag{4.3}$$

The existence of correlations makes it impossible to recover the whole information on $g(\mathbf{z})$, starting from the knowledge of all the Lyapunov exponents.

5. Entropies

A third approach to deterministic chaos passes through the introduction of generalized entropies. In this case we are interested in the information required to characterize a whole trajectory in phase space, rather than the equilibrium distribution. In other words, we introduce the probability $p(\varepsilon; t)$ to detect a trajectory remaining closer than ε to a given orbit for a time t. Analogously to (3.3) we expect that,

$$\left\langle \frac{p(\varepsilon; t)^{q-1}}{\varepsilon^{\tau(q)}} \right\rangle \simeq \exp[-(q-1)K(q)t] \tag{5.1}$$

for $\varepsilon \to 0$ and $t \to \infty$. For $q = 1$ we obtain the metric (or Kolmogorov-Sinai) entropy whose definition is based on the standard Shannon information. For $q = 0$ it is instead receovered the topological entropy which expresses the growth rate of the number of unstable periodic orbits hidden in the strange attractor. Again, as for dimensions, Eq. (5.1) can be simplified, by referring to a fixed grid in phase space, which is the essence of computing entropies from symbol sequences. More precisely, it is convenient to consider symbol sequences obtained from generating partitions. By definition, they are those partitions for which the limit $\varepsilon \to 0$ is not necessary, since it is implied by the limit $t \to$

∞: arbitrarily fine partitions are obtained by interesecting sufficiently many images and preimages of the generator. Given a symbol sequence $\{\ldots, s_{t-1}, s_t, s_{t+1}, \ldots\}$, we indicate with S_t a finite subsequence of legnth t, and call $P\{S_t\}$ the probability to observe S_t at a random position in the infinite string. Eq. (5.1) reads as

$$\sum_{\{S_t\}} p\{S_t\}^q \simeq e^{(1-q)K(q)t} \tag{5.2}$$

where the sum is performed over all subsequences S_t arising from a generating partition. It is obvious that such an approach to the problem can also be formulated in terms of a distribution of "local" entropies. As it is only matter of technicalities we refer to [14] for a more exhaustive discussion.

Before ending the discussion on generalized entropies it is worth to spend a few words to clarify the concept of generating partition, since it plays a central role in connection with thermodynamic formalism. In the last part of Section 3 we have seen how it is possible to map the fractal dimension formalism onto the statistical mechanics problem of a 1-d spin chain. This was a simple task in the case of the Cantor-like set there considered, as a sequence of symbols is straightforwardly associated with each box. More in general, the problem of transforming a dynamical orbit into a sequence of symbols is a typical problem associated with the search for a generating partition, i.e. a partition such that different initial conditions correspond to different symbol sequences. One such partition is immediately found for the logistic map, by splitting the unit interval into two parts lying on the left and on the right of the maximum, respectively [20]. However, passing to two-dimensional maps the problem is much less trivial. Grassberger and Kantz [21] proposed to consider the front composed of the primary homoclinic tangencies [22], as the line dividing the phase space into the elements of a generating partition. There is a growing numerical evidence that their idea is correct, although it has not been yet rigorously proved.

6. Lyapunov exponents, metric entropies and fractal dimensions

To derive relations between Lyapunov exponents and the other chaos indicators one can follow two alternative approaches: a global one based on generalized averages, or a local one, based on the corresponding spectra of such quantities. Here we summarize the results of the first approach. We first discuss metric entropies starting from Eq. (5.1), generalized to cylinders with elliptic cross sections and written as

$$\left\langle \frac{p(\varepsilon;t)^{q-1}}{\varepsilon^{\eta\cdot\tau(q)}} \right\rangle \simeq \exp[(1-q)K(q)t] \tag{6.1}$$

where ε is a vector with components $\varepsilon^{\eta_1}, \varepsilon^{\eta_2}, \ldots$, and $\tau(q)$ is a vector with components $(q-1)D_1(q)$, $(q-1)D_2(q)$, ... The average is taken with respect to the measure μ. The trajectories remaining for a time t in such cylinders around $\mathbf{x(t')}$, are those that started off in an ellipsoid centered on $x(0)$ and characterized by η' with

$$\eta'_j = \begin{cases} \eta_j - t\lambda_j/\ln\varepsilon & \text{if} \quad \lambda_j > 0 \\ \eta_j & \quad\quad\quad \textit{otherwise} \end{cases} \tag{6.2}$$

The probability $p(\varepsilon; t)$ is then equal to the mass $p(\varepsilon'; 0)$ contained in the initial ellipsoid. By inserting Eq. (6.2) into Eq. (6.1), we obtain a statement about the joint distribution of pointwise dimensions and effective Lyapunov exponents

$$\left\langle \frac{p(\varepsilon'; 0)^{q-1}}{\varepsilon^{\eta \cdot \tau(q)}} \varepsilon^{(\eta' - \eta) \cdot \tau(q)} \right\rangle \simeq e^{(1-q)K(q)t} \tag{6.3}$$

For hyperbolic strange attractors, the pointwise dimension along the expanding directions is not a fluctuating quantity (=1), so that the first term in angular brackets can factored out. From the definition of dimension it is equal to 1, and we obtain [25]

$$(q-1)K(q) = -g(1-q, \ldots, 1-q, 0, \ldots, 0) \tag{6.4}$$

where the nonvanishing arguments correspond to the unstable directions. For $q = 1$ Eq. (6.4) reduces to the well known Pesin formula [23]

$$k(1) = \sum_{j=1}^{j^+} \lambda_j D_j(1) \tag{6.5}$$

where j^+ is the number of positive exponents. To discuss the relation between Lyapunov exponents and fractal dimensions, we start from Eq. (3.3) generalized to take elliptic boxes

$$\left\langle \frac{p(\varepsilon)^{q-1}}{\varepsilon^{\eta \cdot \tau(q)}} \right\rangle \simeq 1 \tag{6.6}$$

After n iterations, an ellipsoid with axes centered at $\mathbf{x}$ will be transformed into an ellipsoid with the same mass but with axes

$$\varepsilon'_j = \varepsilon^{\eta'_j} = \varepsilon^{\eta_j} \cdot e^{n\lambda_j} \tag{6.7}$$

By assuming that the evolution is invertible (this is not the case of the logistic map) and substituting Eq. (6.7) in Eq. (6.6), we obtain (taking also into account the invariance of the measure μ under backaward iterations)

$$\left\langle \frac{p(\varepsilon)^{q-1}}{\varepsilon^{\eta \cdot \tau(q)}} \exp[-n\lambda \cdot \tau(q)] \right\rangle \simeq 1 \tag{6.8}$$

This relation can be interpreted as a constraint on the partial dimensions $\mathbf{D}(q)$ which has to be satisfied in order to conserve the average generalized volume given by Eq. (6.6). For sufficiently large n we can assume that the pointwise dimensions (hidden in the masses) are decorrelated from the local effective Lyapunov exponents. Accordingly, the average factorizes and we obtain the simple relation

$$g(-\tau(q)) = 0 \tag{6.9}$$

This is a single equation relating many different variables (partial dimensions D_j). Under the assumption that there is a j_c such that, for $j < j_c$, $D_j = 1$, while, for

$j > j_c$, $D_j = 0$ and, finally, $0 < D_{j_c} < 1$, Eq. (6.9) allows to determine the only remaining unknown quantity D_{j_c}. In the limit case $q = 1$ Eq. (6.9) reduces to the famous Kaplan-Yorke relation [24]

$$D(1) = j_c + \frac{\sum_{j=1}^{j_c} \lambda_j}{|\lambda_{j_c+1}|} \tag{6.10}$$

where j_c is defined as the largest integer such that the sum in the r.h.s. is positive. The idea behind Eq. (6.10) is that all expanding directions contribute with one full unity to the dimension (precisely because of the expansion), the same is assumed to hold for those contracting directions capable to "absorb" such an expansion. The dissipativity of the model ensures the existence of a critical j_c where the expansion turns into a contraction. Another crucial assumption hidden in Eq. (6.10) is that the expansion generated by the unstable manifold is folded primarily along the least unstable directions.Numerical simulations have so far confirmed this conjecture.

If we expect Eq. (6.9) to hold for hyperbolic systems, where it is possible to locally factorize the attractor into the product of stable and unstable manifolds, the same cannot be everywhere true for realistic non-hyperbolic attractor. In Ref. [14] Grassberger, Badii and Politi conjectured that there exists a critical q-value q_c below which the existence of the homoclinic tangencies does not affect the global statistical properties of an attractor, so that we expect Eq. (6.9) still to hold. On the contrary, for $q > q_c$ the scaling behavior of the mass around such points prevails, thus requiring a a different thermodynamic description.

7. The inverse problem

We end the discussion on low dimensional chaos with a sketchy discussion on the possibility to determine the chaotic indicators starting from experimental data. In the previous Sectons we have implicitely assumed that fractal dimension, Lyapunov exponents and metric entropies were computable quantities for a generic dynamical system. However, this is not a trivial task already when the the equations of motion are given (as it is the case of numerical simulations) [25]. In the case of experimental data the problem is even worse. In many cases it is not possible to measure simultaneously many independent variables, to provide a complete characterization of the temporal evolution in phase space. Fortunately, the embedding procedure allows to overcome such a difficulty, even when only a single variable $x(t)$ is available. Indeed, starting from $x(t)$, we can formally define the coordinates

$$y_j(t) = x\left(t + (j-1)\tau\right) \tag{7.1}$$

where τ is a suitable delay. If the underlying dynamics is low dimensional, a suitable number of coordinates $y_j(t)$ can be sufficient to reconstruct the evolution in phase space [26]. More precisely, once the number of variables has been fixed equal to E, the experimental signal is automatically embedded in an E-dimensional phase space, where it is possible to apply the standard techniques to compute, for instance, the fractal dimension. Now, the theorem proved by Takens [26] guarantees that, for E sufficiently large, the embedding technique guarantees a faithful reconstruction of the attractor, where faithful means that there exists a diffeomorphism relating the coordinates $y_j(t)$ with the "true" variables contributing to define the model equations.

This is a very important point, as the possibility to reconstruct the attractor in a suitable copy of the phase space, permits to export the numerical techniques initially developed for numerical simulations. While referring refer to [25] for a more detailed presentation of the various numerical methods, here we list some of the problems connected with a reliable application of the embedding procedure: (a) choice of the embedding delay τ, (b) choice of the sampling time, (c) choice of the number of points (d) filtering of the signal to reduce the experimental error.

References

[1] E.N. Lorentz, J. Atmos. Sci. **20** 130 (1963)

[2] H. Haken, *Advanced Synergetics*, (Springer, Berlin , 1983)

[3] J. Carr, *Applications of Centre Manifold Theory* i (Springer, Berlin, 1981)

[4] S.Smale, Bull. Am. Math. Soc., **73** 747 (1967)

[5] J.P. Eckmann and D. Ruelle, Rev. Mod. Phys. **67**, 617 (1985)

[6] V. Jacobson, Commun Math. Phys. **81**, 39 (1981); M. Misuriewicz, Publ. Math. IHES, **53** 17 (1981)

[7] S. Newhouse, Lectures on dynamical systems in, it Dynamical Systems (Birckhauser, Bosoton 1980)

[8] M. Benedicks and L. Carlson, 1989 *9-th Int. Congr. on Mathematical Physics* ed. B. Simon et al. (Adam Hilger, Bristol) p. 489

[9] B.B. Mandelbrot, *The fractal geometry of nature* (Freeman, San Francisco 1982)

[10] P. Grassberger, Phys. Lett. bf 107A 101 (1985)

[11] P. Grassberger and I. Procaccia, Phys. Rev. Lett. **50** 346 (1983)

[12] R. Badii and A. Politi, Phys. Rev. Lett. **52** 1661 (1984)

[13] R. Benzi, G. Paladin, G. Parisi, and A. Vulpiani, J. Phys. A **17** 3521 (1985); T.C. Halsey, M.H. Jensen, L.P. Kadanoff, I. Procaccia, and B. Shraiman, Phys. Rev. A **33** 1141 (1986)

[14] P. Grassberger, R. Badii, and A. Politi, J. Stat. Phys. **51** 135 (1988)

[15] P. Paoli, A. Politi, and R. Badii, Physica D **36** 263 (1989)

[16] A. Politi, R. Badii, and P. Grassberger, J. Phys. A **21** L763 (1989)

[17] D. Ruelle, *Thermodynamic Formalism* (Addison-Wesley, Reading, Massachusetts 1978); O.E. Lanford, Entropy and Equilibrium States in Classical and Statistical Mechanics in *Statistical Mechanics and Mathematical Problems*, A. Lenard ed. (Springer 1976)

[18] R. Badii, Riv. Nuov. Cim. **12** N3, 1 (1989)

[19] H. Fujisaka, Prog. Theor. Phys. *70* 1264 (1983);

[20] P. Collet and J.-P. Eckmann, *Iterated Maps on the Interval as Dynamical Systems*, Birkhäuser Boston 1980

[21] P. Grassberger and H. Kantz, Phys. Lett. **113A** 235 (1985)

[22] P. Grassberger, in *Chaos in Astrophysics*, J. Perdang et al. (Reidl, Dortrecht, 1985).
[23] Ya. B. Pesin, Russ. Math. Surv. **32**, 55 (1977).
[24] P. Frederikson, J.L. Kaplan, E.D. Yorker, and J.A. Yorke, J. Diff. Eqs. **49** 185 (1983)
[25] R. Badii and A. Politi, *Strange attractors: estimating the complexity of chaotic signals*, in Instabilities and Chaos in Quantum Optics II, ed. by N.B. Abraham et al. (Plenum, New York 1988), p. 335
[26] F. Takens, *Invariants Related to Dimensions and Entropies*, in Atas do 13 Coloquio Brasileiro de Matematica (1984)

ATTENUATION IN FIBERGUIDES AND LYAPUNOV EXPONENTS

Ricardo Lima
Centre de Physique Théorique
C.N.R.S. Luminy Case 907
F - 13288 - Marseille Cedex 09

1. Introduction.

The aim of this paper is to describe some features of simple models for optical waveguides devices. We follow the paper of F. Bentosella and P. Piccoli [1] for the first 3 sections.

Let recall that a coaxial optical fibber consists on a core with refractive indice n_1 and a cladding around the core with a refractive indice n_2 such that $n_2 < n_1$. The radius L_1 of the core is of the order of the micrometer and the radius L_2 of the cladding is about 10-100 times L_1. Instead, the length of such a fibber is of the order of the kilometre.

In principle the device is made in order to be able to propagate an electromagnetic field with infinite many modes in the cladding and only a finite number, even a single one, mode in the core.

Since $n_2 < n_1$, in absence of imperfections, the wave would propagate along the axis at infinite distances and, in particular the information carried by the core will be transmitted at distances without any power lost.

But in particular, because imperfection in the size and materials homogeneity of the guide can not be avoided, we observe an attenuation of the out put intensity with respect to the corresponding input.

In order to measure this effect an attenuation per unit length is defined as :

$$\beta = -10 \ L^{-1} \log_{10} \frac{I_{out}}{I_{in}} \qquad (1.1)$$

where I_{out} and I_{in} are the output and input intensities and L is the length of the guide. This is certainly a mean quantity and should be measured in a certain range of values of L.

As it is clear (1.1) is related to the rate of an exponential decay of amplitudes of the waves which propagates inside the guide.

This fact brings us from the study of the attenuation of the propagation along wave guides to the so called localization problem which treats the exponential decay of solutions of differential equations.

But, before going to the equation let's take a glance at the different kind of imperfections which can cause the localization of the waves.

Let x be the axial direction and (y, z) the coordinates for the plane orthogonal to the axial direction.

Then, in some cases the imperfections are due to fluctuations of the values of the indexes n_1 and n_2 along the x-axis and therefore this leads us to introduce functions $n_1(x)$ and $n_2(x)$ in the equations. It can also be that the radius of the core is x-dependent and therefore these forces to introduce a position - dependent refractive index

$$\begin{aligned} n(x, y, z) &= n_1 \text{ if } \sqrt{(y^2 + z^2)} \leq L_1(x) \\ &= n_2 \text{ if } L_1(x) \leq \sqrt{y^2 + z^2} \leq L_2 \end{aligned} \tag{1.2}$$

Finally, it may happen that mechanical deformations of the axis of the fibber imply that the refractive index became x-dependent via an implicit formula such as :

$$n(x, y, z) = n(x, y - y_0(x), z - z_0(x)) \tag{1.3}$$

where $(y_0(x), z_0(x))$ in the position of the centre of the fibber in the plane orthogonal to the x-direction at place x.

Now, in any case, Maxwell's equations for a dielectric give rise to the following equations for the electrical field :

$$\nabla^2 \vec{E}(x, y, z) + \vec{\nabla} \cdot \left[\frac{\vec{E}(x, y, z) \cdot \vec{\nabla} \cdot n^2}{n^2} \right] + k^2 n^2 \vec{E}(x, y, z) = 0 \tag{1.4}$$

which is fulfiled for the amplitude of a wave of the form :

$$\vec{E}(x, y, z) \; e^{i \omega t} \tag{1.5}$$

with

$$k^2 = \mu_0 \varepsilon_0 \omega^2 \tag{1.6}$$

And, up to first order in the variation of n, we get

$$\nabla^2 \vec{E}(x, y, z) + k^2 n^2 (x, y, z) \vec{E}(x, y, z) = 0 \tag{1.7}$$

which is the usual "scalar" equation.

The same equation is also to be verified by the magnetic field with the only difference on the boundary conditions.

Therefore, any component of the field is a solution of an equation like (1.7) :

$$- \Delta \psi(x, y, z) - \frac{\omega^2}{C^2} n^2(x, y, z) \psi(x, y, z) = 0 \tag{1.8}$$

where ψ is to be considered as a vector of the Hilbert space $L^2(R \times D)$, D being a disc, and the domain of the operator Δ is fixed by the boundary conditions, according to which composant of the field we treat.

Now in (1.8) the operator

$$- \Delta - \frac{\omega^2}{C^2} n^2(x, y, z) \tag{1.9}$$

can be decomposed in an integral along the field :

$$x \to \mathcal{H}_x = L^2(D) \tag{1.10}$$

and therefore we end up with a field of operators

$$x \to H_x \tag{1.11}$$

where $H_x : \mathcal{H}_x \to \mathcal{H}_x$ is defined by

$$H_x = - \left(\frac{d^2}{dy^2} + \frac{d^2}{dz^2} \right) - \frac{\omega^2}{C^2} n_x(y, z) \tag{1.12}$$

It is natural to express the solution ψ of the global problem (1.8) as a field

$$x \to \psi_x(y, z) \tag{1.13}$$

and for each x to write ψ_x as a linear combination of the eigenfunction of (1.12).

For a circular and concentric shape, i.e if $n_x(y, z)$ is only radial in each x (but not necessarily the same function for different x), the eigenfunctions of (1.12) are known to be the Bessel functions $\phi_i^{(x)}(y, z)$, i = 1,..., the x. label being for notational purposes to be clear in a moment.

Notice that (1.12) is a Schrödinger type operator with a potential given by n_x.

If we denote :

$$a_j(x) = \left(\phi_j^{(x)}, \psi_x\right) \tag{1.14}$$

where ψ_x is as in (1.13) and the scalar product is taken in $\mathcal{H}_x$, we can write :

$$\psi_x(y, z) = \sum_{j=1}^{\infty} a_j(x)\ \phi_j^{(x)} \tag{1.15}$$

but the equations for the $a_j(x)$, coming from (1.8) are an infinite-set of coupled equations since the operator couple the different a_j.

So, it seems that this way of seeing through the problem doesn't help us very to much...

One way to get out of this difficulty is .... to consider only one mode !

2. Decoupled Equations

In this case the linear equation reads :

$$-\frac{d^2}{dx^2} a_j(x) - 2 K_{jj}(x) \frac{da_j(x)}{dx} - L_{jj}(x)\, a_j(x) + E_j^{(x)} a_j(x) = 0 \tag{2.1}$$

with

$$K_{jj}(x) = \left(\phi_j^{(x)} \mid \frac{d}{dx}\phi_j^{(x)}\right) \tag{2.2}$$

and

$$L_{jj}(x) = \left(\phi_j^{(x)} \mid \frac{d^2}{dx^2}\phi_j^{(x)}\right) \tag{2.3}$$

(2.1) is to be fulfilled for each j. By a classical change of variables the first derivative of a_j in (2.1) can be eliminated, and the resulting equation is :

$$\frac{d^2}{dx^2} b(x) + \left(-\frac{dK(x)}{dx} - K(x)^2 + L(x) - E^{(x)}\right) b(x) = 0 \tag{2.4}$$

where the index j is omitted for simplicity of the notation.

Now we can consider the x-dependence of n_x and therefore of K(x) and L(x) to be a random ; in such a case (2.4) became the equation for a random oscillator.

Dissipation will introduce some correlation in the "noise" therm :

$$W_x = -\frac{dK(x)}{dx} - K(x)^2 + L(x) \tag{2.5}$$

which needs to be taken in account.

If we write

$$W_x = \langle E^{(x)} \rangle + F_x \tag{2.6}$$

the remarkable fact is the dependence of F_x in the derivative of K, which in turn depends on the derivative of the position of the centre band and then can not be neglected.

A result of Arnold, Papanicolaou and Wihstutz for random oscillators give the exponential decay γ of the solution of (2.4) :

$$|a(x)| \sim e^{-\gamma|x|} \tag{2.7}$$

In the correlation of F_x is given by

$$C(u) = < F_{x+u}\ F_x > - < F_x > < F_{x+u} >$$

$$= \alpha^2 e^{-u^2/L_c^2} \tag{2.8}$$

then, these authors give

$$\gamma \cong \alpha^4\ L_c < n > \Delta n^2 e^{-cst\ n^2 L^2} \tag{2.9}$$

and, in particular,

$$\gamma > 0 \tag{2.10}$$

therefore this leads to the somewhat unexpected result that, even without coupling, the amplitude of guide modes is always localized.

3. From Wave guides to Transfer Matrices.

Another possible treatment of the problem is to made a space discretization in x-direction and to use radial symmetry in y - z plane. In that case we suppose that the waves will freely propagate between each two sections distant from each other by a constant mesh δ.
Or, equivalently, the changes of n_x are made for $x = q\delta$, $q = ..., -1, 0, 1, ...$.

If we consider the problem of N modes, the discrete (in x) analogue of (1.8), using the same basis as in the previous section, will bring to an equation for a N-dimension vector $\vec{A}(q)$, for each section q. The components of $\vec{A}(q)$ being the N modes. Using the equivalence of this second order equations to a system of the first order equations we can rewrite the relation between the $\vec{A}(q+1)$, $\vec{A}(q)$ and $\vec{A}(q-1)$ in the form of a transfer matrice equation :

$$\vec{B}(q+1) = T_q\ \vec{B}(q) \tag{3.1}$$

Here $\vec{B}(q)$ is the 2N - dimensional vector the first N components being those of $\vec{A}(q)$ and the last N components those of $\vec{A}(q-1)$; T_q is a 2N x 2N matrice defined for each q by

$$T_q = \begin{bmatrix} K^{-1}(q, q+1) \cdot D(q) & -K(q, q+1)\, K(q, q-1) \\ 1 & 0 \end{bmatrix} \qquad (3.2)$$

where K(q, q+1), D(q), 1 and 0 are N x N matrices defined by

$$\begin{aligned} K_{ij}(q, q+1) &= \left(\varphi_i^{(q)} \mid \varphi_j^{(q+1)}\right) \\ D_{ij}(l) &= \left(2 + \delta^2 E_i^{(q)}\right) \delta_{ij} \\ 1_{ij} &= \delta_{ij} \\ 0_{ij} &= 0 \end{aligned} \qquad (3.2)$$

The reader should observe that (3.1) with the help of (3.2) is equivalent to the original equation :

$$\vec{A}(q+1) = K^{-1}(q, q+1)\, D(q)\, \vec{A}(q) - K(q, q+1)\ K(q, q-1)\ \vec{A}(q-1) \qquad (3.4)$$

It can be shown that equation (3.1) has the same solution (i.e. generates the same wave) as a Schrödinger equation with a double circular well potential of radius L_2 and L_1 and depth $-K_2^2$ and $-K_1^2$ and therefore the properties of the solutions $\vec{A}(q)$, the eigenvectors, will depend of the "potential" K and D.

In order to arrive at the context if symplectic transfer matrices we may notice that invertible changes of basis G_q will transform T_q in a symplectic matrice :

$$S_q = G_q^{-1}\ T_q\ G_{q-1} \qquad (3.5)$$

which can be written explicitly :

$$S_{q+1} = \begin{bmatrix} J_q^{*} D(q)\, J_q & -1 \\ 1 & 0 \end{bmatrix} \qquad (3.6)$$

with

$$J_q = K(q, q-1) \cdot K(q-1, q-2) \ldots K(1, 0) \qquad (3.7)$$

It turns out then that we can write :

$$\vec{C}(q) = S_q \; S_{q+1} \ldots S_0 \; \vec{C}(0) \tag{3.8}$$

with

$$\vec{C}(q) = G_q \; \vec{B}(q) \tag{3.9}$$

and therefore the assymptotic behaviour (when $q \to \infty$) is governed by the assymptotic properties of the product of matrices

$$P_q = S_q \; S_{q+1} \ldots S_0. \tag{3.10}$$

Suppose, for instance, that

$$||P_n \vec{\xi}|| \sim C.\, e^{\sim \gamma n} \quad \text{as } n \to \infty \tag{3.11}$$

then γ is (up to a constant) the attenuation per unit length defined by (1.1).

Recall that

$$||P_n \vec{\xi}||^2 = \left(P_n \vec{\xi}, P_n \vec{\xi}\right) =$$

$$= \left(\vec{\xi}, P_n^* P_n \vec{\xi} \right) \tag{3.12}$$

where P_n^* is the adjoint matrice of P_n.

4. Attenuation and Lyapunov exponents.

There are several difficulties which came out if we try to give a precise meaning to (3.11).

First it may very well be that γ, as it appears in (3.11), depends on the vector $\vec{\xi}$ since there is no a priori reason for having a uniform decreasing rate on all the vector space.

Second, in our case, P_n is not exactly a matrice, but as well as the S_q, it should be considered as a random field of matrices, in the sense that we have considered the S_q as issues of a random sample. Therefore, P_n is better understood as a product of random matrices.

Last, but not least, the assymptotic behaviour as $n \to -\infty$ should also be taken in account since the starting equation is a two-sided infinite equation and therefore the solutions should be interpreted as $x \to \pm \infty$.

It turns out, that the same problem appeared in the case of the Schrödinger equation.

The two first difficulties are overcome by a strong result, due to Oseledec [2], that we now state in a version due to Ruelle :

Theorem. Let μ be a probability on a space X, and $t : X \to X$ a measure preserving map such that μ is (invariant and) ergodic.

Let

$$S : X \to M(\mathbb{R}^n) \tag{4.1}$$

be a measurable map from X to the set of n x n real matrices, such that

$$\int_X \log^+ ||S_x|| \, d\mu(x) < +\infty \tag{4.2}$$

where $\log^+ u = \max(0, \log u)$.

Then, for almost all $x \in X$ (with respect to the measure μ), the following limit exist and doesn't depend on x :

$$\Lambda_x = \lim_{K \to \infty} \left(P_k(x)^* \, P_k(x)\right)^{\frac{1}{2n}} \tag{4.3}$$

where $P_k(x)$ is the product of the K matrices :

$$P_k(x) = S_{t^{k-1}(x)} \cdot S_{t^{k-2}(x)} \cdots S_{t(x)} \, S_x \tag{4.4}$$

and $P_k(x)^*$ is the adjoint of $P_k(x)$.

Λ_x is a positive matrice and the logarithms of it are called Liapunov exponents. Again the Liapunov exponents $\lambda_1 \geq \lambda_2 \geq \ldots$ are independent of x almost everywhere.

In order to see the exact relation with the previous section we also give a version of the Oseledec theorem which is more suitable in our context (in fact more closed to the original one !) :

Theorem : With the same assumptions of the previous theorem, for each $x \in X$ we can write

$$\mathbb{R}^n \equiv V_x^1 \supset V_x^2 \supset \ldots \supset V_x^n. \tag{4.5}$$

where the V_x^i are vectorial spaces of dimension (n - i + 1) and if $\vec{\xi} \in V_x^i$ and $\vec{\xi} \notin V_x^{i-1}$ we have :

$$\Lambda_i = \lim_{K \to \infty} \frac{1}{K} \log ||P_k(x) \, \vec{\xi}|| \, , \, i = 1, \ldots . \tag{4.6}$$

where the λ_i are the Liapunov exponents.

If we now apply this result to the situation where the S_x are the S_q of the previous section (see(3.6)) are see that, since we are interested in the fast decay of the solutions, this is given by λ_n. For, the S_q and therefore the P_q and Λ are symplectic matrices ; therefore the eigenvalues of Λ are n positive numbers and their inverses that we can write :

$$e^{\lambda_1}, e^{\lambda_2}, \ldots, e^{\lambda_n}, e^{-\lambda_n}, \ldots, e^{-\lambda_2}, e^{-\lambda_1} \tag{4.7}$$

where to each positive Liapunov exponent λ_1 correspond a symmetric one $-\lambda_i$.

Now for the vectors $\vec{\xi}$ in the suitable subsets of $\mathbb{R}^n$ given by the previous theorem we shall have

$$\text{when } k \to +\infty, \quad ||P_k \vec{\xi}|| \sim e^{\pm\lambda_i k} \tag{4.8}$$

From where we can see that the decay of the solution (for $k \to +\infty$) is given by the smallest positive Liapunov exponent λ_n.

There remains the problem of gluing together a solution for $k > 0$ and a solution for $K < 0$. As both are built from the initial vector $\vec{C}(0)$ it is not at all certain that there is a $\vec{C}(0)$ at the same time in $V^n \setminus V^{-n}$ for the product P_k, $k > 0$ and for P_k, $k < 0$.

That such is the case is a consequence of the existence for almost all E of a solution $\vec{\xi}$ of the Schrödinger equation :

$$(H - E)\,\vec{\xi} = 0 \tag{4.9}$$

which is slowly increasing when $k \to \pm\infty$:

$$\lim_{|k| \to \infty} \frac{\xi_k}{k^{1/2+\varepsilon}} = 0 \tag{4.10}$$

Now if $\lambda_n > 0$ and the two subsets $V^n \setminus V^{-n}$ for P_k, $k > 0$ and $k < 0$ have no vector other than 0 in common, all the solutions obtained by applying P_k to any initial condition should exponentially increase at least in one of the two directions ($k > 0$ or $k < 0$) and (4.10) will never be fulfilled.

Therefore the problem of the exponential localization of the solutions of the wave equations as they are given by (3.4) can be solved by the computation of Liapunov exponents for the product of random matrices (3.10) in the infinite limit.

5. Lyapunov Exponents and Models.

Explicit computation of Lyapunov exponents is a very difficult task, even for the simplest models.

There are two possible approaches to deal with the problem in a reasonable way : either we perform numerical computations or, using analytical estimations, we deduce some scalling laws of the Lyapunov exponents as functions of a perturbation parameter of the model.

Next we review a few results in the latter direction, which can be of interest for the localization problem.

In [3] we studied a very simple model of product of matrices which can nevertheless give some insight on the localization lengths as the randomness is changed, i.e when the mean and the fluctuations of the defects in the waveguide vary.

Let's consider two 2 x 2 matrices A and B of the form (3.6) :

$$A = \begin{pmatrix} 2a & -1 \\ 1 & 0 \end{pmatrix} \tag{5.1}$$

$$B = \begin{pmatrix} 2b & -1 \\ 1 & 0 \end{pmatrix} \tag{5.2}$$

where $a, b \in \mathbb{R}$.

A and B are symplectic matrices with determinant one. Each of therm is elliptic or hyperbolic according $|a| < 1$ (resp $|b| < 1$) or not, since we have

$$|\mathrm{Tr}\,(A)| = 2a \tag{5.3}$$

and

$$|\mathrm{Tr}\,(B)| = 2b \tag{5.4}$$

Each of this matrice should be vilued as one of two possible transfer matrices from one section of guide to the following section.

We then have a model for the one-mode problem.

Clearly, if only one of the two matrices A or B is present , i.e. if there is no defect, the Liapunov exponent is 0 or strictly positive according to whether the given matrice is elliptic or hyperbolic. In the first case the norm of any transformed vector by a product of matrices remains bounded and, on the contrary, it will grow exponentially fast in the last case.

Since where no defects are present we wait for a propagation without any power decay, the case of the two elliptic matrices is clearly the better adapted to this situation.

We therefore consider the following simple model (Bernoulli model) :

For each step the transfer matrice is chosen randomly according a Bernouilli law, i.e. A with probability p and B with probability (1 - p) in an independent way from step to step. If we take X to be the set of all double-infinite sequences of 0 and 1, equipped with the product measure given by p and (1 - p) and t to be the shift, we see that Oseledec theorem applies in this case. Therefore Lyapunov exponents exist almost everywhere and are almost everywhere independent of the sample.

Since A and B have determinant 1, the two eigenvalues of Λ (see 4.3) can be written

$$\alpha_{\pm} = \exp\,(\pm\,\lambda) \tag{5.5}$$

and therefore the two Lyapunov exponents are $\pm\,\lambda$.

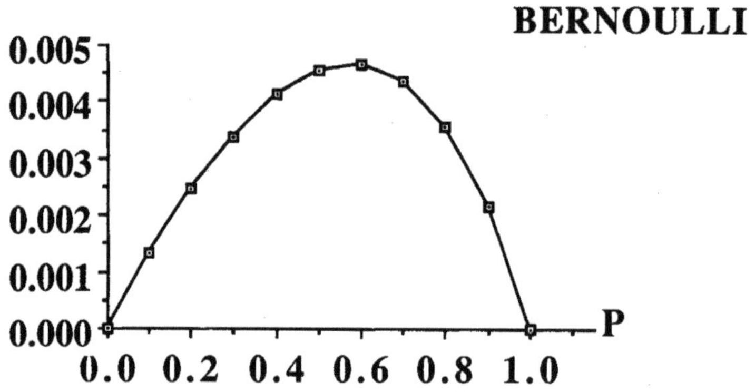

Fig. 1

Fig. 1. Lyapunov exponent as function of the probability p for a Bernoulli model of two elliptic matrices (from [3]).

Fig. 1 shows a plot of the positive Lyapunov exponent $+\lambda$ as a function of the probability p for the case where a = 0,9 and b = 0,8.

The precise analysis of this curve is out of the scope of the present report, the reader is referred to [3] for a more detailed analysis. Notwithstanding it can be observed that the Lyapunov exponent grows as we move from the "perfect" guide (p = 0 or p = 1) but not in a completely symmetric way.

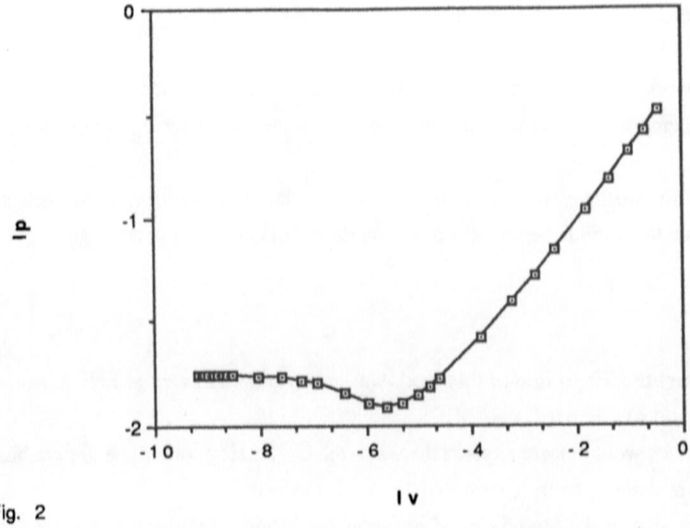

Fig. 2

Fig. 2. Lyapunov exponent as a function of the variance v of the process (log - log scale). (from [4]).

Fig. 2 shows a plot of another numerical experiment taken from [4]. Here, instead we studied a model for which the values of the matrix elements are chosen accordingly a random process for which the mean value and the variance (the second moment) can be independently monitored.

What was proved in [4] and is clearly seen on the plot is a cross-over effect : for small variance the Lyapunov exponent doesn't change very much and it is shown that inside this region it is mean-dependent ; but as soon as some critical value of the variance of the random process is overcame the Lyapunov exponent grows with a power law given by the slape of the (assymptotic) straightline on the right side of the plot (since it is in log-log scale).

Without going into greater detail, we underline the agreement of the previous conclusion with the analysis of section 2 about the dependence of the localization length on the derivative of the transfer function K.

It turns out that fluctuations in the structure of the waveguide can be responsible for the dominant effect in the attenuation of the power transfer in an optical fibber. This is in a quite good agreement with experimental measurements.

6. Conclusion.

The analysis of the attenuation in propagation of an electromagnetic wave in an optical guide can be related to a localization problem for which the use of Lyapunov exponents are useful.

Therefore we have review some of their fundamental properties as well as its dependence upon the various parameters of the randomly disturbances in the regularity of the guide.

Acknowledgements : The origin of this work was an invited talk given at ENEA workshop in Bologna, on NON LINEAR DYNAMICS.

The author would like to express thanks to the Organizing committee to give him the opportunity of formalizing some ideas about the present subject.

Thanks also to the organizers of the long workshop "Evolution and Complexity" (April to July 1990), held at the Institute for Scientific Interchange in Torino during which this report was written.

References :

[1] F. Bentosella, P. Piccoli, J. Phys. (France) **49**, 2001 (1988)

[2] V.I. Oseledec, Trudy Mosk, Mat Obsc **19**, 179 (1968)

[3] R. Lima, P. Piccoli, Japan Journal of Appl. Mathematics, Vol. 7, n° 3 (1990)

[4] R. Lima, M. Rahibe, J. Phys. A : Math. and general, **23**, 781-789 (1990)

Analytical Radiative Corrections to the Anomalous Magnetic Moment of the Electron

by S. Turrini
Dipartimento di Fisica dell'Università di Bologna; Italy
and Istituto Nazionale di Fisica Nucleare, sezione di Bologna; Italy

Summary

§1 PREMISE AND GENERAL PROBLEMS

Electromagnetic vertex for a mass shell electron and a photon of arbitrary momentum Δ is written (*)

$$\mathcal{V}_\mu(p,\Delta) = (-e)\bar{u}_2(p-\frac{1}{2}\Delta)\Big(F_1(-\Delta^2)\gamma_\mu - \frac{i}{4m^2}F_2(-\Delta^2)(\gamma_\mu\not{\Delta}-\not{\Delta}\gamma_\mu)\Big)u_1(p+\frac{1}{2}\Delta) \quad (1.1)$$

where $-e, m$ are the electron charge and mass, $\bar{u}_2, u_1$ the spinor wave function of final and initial electron, and $p \mp \frac{1}{2}\Delta$ their four momenta. $F_1(-\Delta^2)$ and $F_2(-\Delta^2)$ are the Dirac and Pauli form factors, depending only on $-\Delta^2 = \Delta_0^2 - |\vec{\Delta}|^2$, the moment transfer. Renormalized perturbative QED expansion gives them as power series in α/π, α being the fine-structure constant [1]:

$$\begin{aligned} F_1(-\Delta^2) &= 1 + \sum_{n=1}^{\infty}\Big(\frac{\alpha}{\pi}\Big)^n F_1^{(2n)}(-\Delta^2) \\ F_2(-\Delta^2) &= \sum_{n=1}^{\infty}\Big(\frac{\alpha}{\pi}\Big)^n F_2^{(2n)}(-\Delta^2) \end{aligned} \quad (1.2)$$

hence to zeroth order in α one has simply: $F_1(-\Delta^2) = 1 + O(\alpha)$, $F_2(-\Delta^2) = 0 + O(\alpha)$.

(*) Metric is $\delta_{\mu\nu}$ $\mu,\nu = 1,2,3,4$. For instance, an electron on mass shell has momentum $p_\mu = (\vec{p}, p_4) = (\vec{p}, ip_0)$, $p_\mu p_\mu = |\vec{p}|^2 - p_0^2 = -m^2$ and its propagator is: $(-i)\dfrac{-i\not{p}+m}{p^2+m^2}$

Conservation of electric charge requires $F_1(0) = 1$, F_1 being the whole, physical, renormalized Dirac form factor at zero momentum transfer, i.e. $-\Delta^2 = 0$. A practical rule for renormalizing Dirac form factor is to subtract from it its infinite value at zero momentum transfer; the form factor so obtained satisfies $F_1(0) = 1$. On the contrary $F_1(-\Delta^2)$, Pauli form factor, needs no renormalization: it vanishes at zero order, but not at zero momentum transfer, and it is a pure radiative correction.

As it is well known, the anomalous magnetic moment of the electron is:

$$\mu_e = g\frac{1}{2}\mu_B,$$

where μ_B is Bohr magneton.
One has from QED: $g = 2(F_1(0) + F_2(0))$. At zeroth order in α due to (1.2) one has $g = 2$, but, according to radiative corrections and the condition $F_1(0) = 1$, one has

$$g = 2(1 + F_2(0))$$

It is preferred to define the "electron anomaly" a_e so that

$$a_e \equiv F_2(0) = \frac{1}{2}(g - 2)$$

From (1.2) one has then that also the "electron anomaly" a_e is expressed as a power series in α/π :

$$a_e^{th} = \sum_{n=1}^{\infty} a_n(\frac{\alpha}{\pi})^n \tag{1.3}$$

§2 EXPERIMENTAL AND THEORETICAL VALUES

a) Experimental Values:

The anomalous magnetic moment of the electron and of the positron have been measured up to 0.004 p.p.m. [2],[3]:

$$\begin{aligned} a_{e^-}^{exp} &= 1\ 159\ 652\ 188.4\ (4.3) \times 10^{-12} \\ a_{e^+}^{exp} &= 1\ 159\ 652\ 187.9\ (4.3) \times 10^{-12} \end{aligned} \tag{2.1}$$

Experimental error in (2.1) derives from several sources:

statistical error :	0.62×10^{-12}
error due to microwave power shift :	1.3×10^{-12}
error due to cavity shift :	4×10^{-12}

It is worth while to note that the equivalence of the values for e^- and $e+$ is one of the best proofs of charge conjugation. Such an accuracy demands comparable or better theoretical precision to become perhaps the most precise test of QED, or the QED definition of α (which would be now the most accurate) to be compared with the determinations offered by the new solid–state effects.

b) Theoretical Values:

In perturbative QED a_e is written (1.3):

$$a_e^{th} = a_1(\frac{\alpha}{\pi}) + a_2(\frac{\alpha}{\pi})^2 + a_3(\frac{\alpha}{\pi})^3 + a_4(\frac{\alpha}{\pi})^4 + \dots$$

where $a_n, n = 1, 2, \dots$ are numerical coefficients free from infrared divergencies. First and second order coefficients, a_1 and a_2, have been known analytically since a long time [1],[4]:

$$\begin{aligned} a_1 &= \frac{1}{2} \\ a_2 &= \frac{197}{144} + \frac{1}{2}\zeta(2) + \frac{3}{4}\zeta(3) - 3\zeta(2)\lg(2) + \frac{1}{45}(\frac{m_e}{m_\mu})^2 = \\ &= -0.328\,478\,965 + 0.000\,000\,519\,8 \; = -0.328\,478\,445 \end{aligned} \tag{2.2}$$

[the contribution $+\frac{1}{45}(\frac{m_e}{m_\mu})^2 = +0.0000005198$ derives from muon internal loops].

Riemann $\zeta(n)$ function is used: $\zeta(n) = \sum_{p=1}^{\infty} \frac{1}{p^n}, \quad n = 2, 3, \dots$
whose values are (analytical values are known only for even integers arguments):

$$\begin{aligned} &\zeta(2) = 1.644\,934\,067 \dots = \frac{\pi^2}{6} && \zeta(3) = 1.202\,056\,903 \dots \\ &\zeta(4) = 1.082\,323\,234 \dots = \frac{\pi^4}{90} && \zeta(5) = 1.036\,927\,755 \dots \\ &\dots \\ &\zeta(2n) = \text{const } \pi^{2n} \end{aligned}$$

The value of third and fourth order coefficients a_3 and a_4 is known only numerically [3]:

$$\begin{aligned} a_3 &= 1.175\,62\,(56) && \text{(error 0.5\%)} \\ a_4 &= -1.472\,(152) && \text{(error 10 \%)} \end{aligned} \tag{2.3}$$

Third order coefficient a_3 consists of 72 Feynman graphs (40 of them different) belonging to 3 different sets:

1) 6 *light light* graphs, known only numerically [3],[8],

2) 16 *vacuum polarization* graphs, known analytically [5],

3) 50 *three photon exchange* graphs ([3],[6],[7],[8] and references therein) subdivided into three topological families:

3a) twice reducible diagrams (no crossed photon lines), known analytically

3b) once reducible diagrams (2 crossed photon lines):
some are known analytically, but the whole family is known only numerically.

3c) irreducible diagrams (3 crossed photon lines), known only numerically.

Fourth order coefficient a_4 consists of 891 Feynman graphs, and its calculation can be done only numerically, not only due to the number of the graphs, but also for the essentially elliptical nature of required integrals [3]. Some simple graphs have been calculated analytically mainly to check numerical calculation of the whole [9].

It seems impossible as far to now to calculate higher order coefficients $a_n, n \geq 5$ also numerically, nevertheless some simple graphs have been calculated (analytically or only numerically) [9].

Other contributions to electron anomaly are [3]:

$$a(\mu \text{ meson}) = +\frac{1}{45}\left(\frac{m_e}{m_\mu}\right)^2\left(\frac{\alpha}{\pi}\right)^2 \simeq 2.8 \times 10^{-12}$$

$$a(\tau \text{ meson}) \simeq 0.01 \times 10^{-12}$$

$$a(\text{hadrons}) \simeq 1.6(2) \times 10^{-12}$$

$$a(\text{weak}) \simeq 0.05 \times 10^{-12}$$

Muon contribution has been already included in (2.2). The importance of the hadronic contribution has to be noted.

Summing up the results (2.2) (2.3) and using the most precise available value of α (from QHE, Quantum Hall Effect) [10]:

$$\alpha^{-1}(\text{QHE}) = 137.035\ 997\ 9\ (33) \qquad (\text{error } 0.024 \text{ p.p.m.}) \tag{2.4}$$

the theoretical value of a_{e^-} becomes:

$$a_{e^-}^{th} = 1\ 159\ 652\ 133\ (29) \quad \times 10^{-12} \tag{2.5}$$

which agree within 2 standard deviation with (2.1) :

$$a_{e^-}^{exp} = 1\ 159\ 652\ 188.4\ (4.3) \times 10^{-12}$$

The error in the theoretical value (2.5) of a_{e^-} comes [3]:

$$7.1 \times 10^{-12} \text{ from } a_3$$

$$4.5 \times 10^{-12} \text{ from } a_4$$

$$28. \times 10^{-12} \text{ from } \alpha$$

The error is still dominated by the error of α, so to have a better test of QED it is necessary to improve the value of α. On the other hand it is possible to obtain the value of α comparing the experimental and theoretical values of a_e to get:

$$\alpha^{-1}(\text{QED}) = 137.035\,991\,4\ (11) \qquad (\text{error } 0.0081 \text{ p.p.m.}) \tag{2.6}$$

which is better than previous non QED results (2.4). Errors derive [3]:

0.0037 p.p.m. from experiment
0.0072 p.p.m. from theory.

It is worth to note that the theoretical error is larger then the experimental error: it becomes then necessary to improve the theoretical value!

§3 TECHNIQUES AND ALGORITHMS

–3a) Form factor decomposition

To calculate electron anomaly it is enough to know the off mass shell vertex amplitude up to the first order in the photon momentum and to expand it with respect to Δ for an arbitrary p [6]:

$$\mathcal{W}_\mu(p,\Delta) = \mathcal{V}_\mu(p) + \Delta_\nu \mathcal{T}_{\mu\nu} + \dots \tag{3a.1}$$

On invariance grounds the following decompositions hold:

$$\begin{aligned}
\mathcal{V}_\mu(p) &= V_1\gamma_\mu + V_2\frac{1}{2}\{\gamma_\mu, i\not{p}+1\} + V_3 i p_\mu(i\not{p}+1) + V_4\frac{1}{2}[\gamma_\mu, i\not{p}+1] \\
\mathcal{T}_{\mu\nu}(p) &= T_1\frac{1}{2}i[\gamma_\mu,\gamma_\nu] + T_2\frac{1}{4}\{[\gamma_\mu,\gamma_\nu],\not{p}\} + T_3 i[p_\mu,\gamma_\nu]\not{p} + T_4\frac{1}{2}i[\{\not{p}_\mu,\gamma_\nu\},\not{p}] \\
&\quad + T_5 i\delta_{\mu\nu} + T_6\delta_{\mu\nu}\not{p} + T_7 i p_\mu p_\nu + T_8 p_\mu p_\nu \not{p} + T_9[p_\mu,\gamma_\nu] + T_{10}\{p_\mu,\gamma_\nu\}
\end{aligned} \tag{3a.2}$$

where $[A,B] \equiv AB - BA$, $\{A,B\} \equiv AB + BA$, and electron mass $= 1$ everywhere. Having already taken the $\Delta \to 0$ limit, all the above factors depend only on the variable $u \equiv -p^2$:

$$\begin{aligned}
V_i &\equiv V_i(u) \qquad & i &= 1,\dots,4 \\
T_j &\equiv T_j(u) \qquad & j &= 1,\dots,10
\end{aligned} \tag{3a.3}$$

Moreover, due to charge conjugation, $V_4 = 0$ and $T_j = 0, j \geq 5$.
Closing $\mathcal{W}_\mu$ between spinor states and comparing with usual expression for the vertex in the case of mass–shell electrons and arbitrary photon momentum (1.1), one finds that the anomaly is:

$$a \equiv F_2(0) = V_2(1) - 2T_1(1) - 2T_2(1) \tag{3a.4}$$

To obtain V_i, T_j to be used in (3a.4) one has to work out, by a straighforward but cumbersome calculation, projectors $P^i_\mu(p), P^j_{\mu\nu}(p)$ to obtain form factors by a suitable trace:

$$\begin{aligned} V_i &= Tr\ \{P^i_\mu(p)\mathcal{V}_\mu(p)\} \qquad i=1,\ldots,4; \\ T_j &= Tr\ \{P^j_{\mu\nu}(p)\mathcal{T}_{\mu\nu}(p)\} \qquad j=1,\ldots,10 \end{aligned} \tag{3a.5}$$

These projectors are found to be:

Vectorial Projectors for $\mathcal{V}_\mu(p)$:

$$\begin{aligned} P^1_\mu(p) &= \frac{1}{12p^2}\left((p^2+1)\gamma_\mu + 3ip_\mu - (\frac{4}{p^2}+1)p_\mu \not{p}\right) \\ P^2_\mu(p) &= \frac{1}{12p^2}\left(\frac{4}{p^2}p_\mu\not{p} - 3ip_\mu - \gamma_\mu\right) \\ P^3_\mu(p) &= \frac{1}{12p^2}\left(\gamma_\mu - \frac{4}{p^2}p_\mu\not{p}\right) \\ P^4_\mu(p) &= \frac{i}{12p^2}\left(p_\mu - \not{p}\gamma_\mu\right) \end{aligned} \tag{3a.6}$$

Tensorial Projectors for $\mathcal{T}_{\mu\nu}(p)$:

$$\begin{aligned} P^1_{\mu\nu}(p) &= \frac{-i}{24p^2}\left(-\frac{1}{2}p^2[\gamma_\mu,\gamma_\nu] + \not{p}[p_\mu,\gamma_\nu]\right) \\ P^2_{\mu\nu}(p) &= \frac{1}{24p^2}\left(-\frac{1}{2}\not{p}[\gamma_\mu,\gamma_\nu] + [p_\mu,\gamma_\nu]\right) \\ P^3_{\mu\nu}(p) &= \frac{-i}{24p^2}\left(-\frac{1}{2}p^2[\gamma_\mu,\gamma_\nu] + 2\not{p}[p_\mu,\gamma_\nu]\right) \\ P^4_{\mu\nu}(p) &= \frac{-i}{24p^2}\left(-2p_\mu p_\nu + \not{p}\{p_\mu,\gamma_\nu\}\right) \\ P^5_{\mu\nu}(p) &= \frac{-i}{12p^2}\left(p^2\delta_{\mu\nu} - p_\mu p_\nu\right) \\ P^6_{\mu\nu}(p) &= \frac{1}{12p^4}\left(p^2\delta_{\mu\nu} + p_\mu p_\nu\right)\not{p} \\ P^7_{\mu\nu}(p) &= \frac{-i}{12p^4}\left(-p^2\delta_{\mu\nu} + 4p_\mu p_\nu\right) \\ P^8_{\mu\nu}(p) &= \frac{1}{12p^6}\left(-p^2\not{p}\delta_{\mu\nu} + 6\not{p}p_\mu p_\nu - p^2[p_\mu,\gamma_\nu]\right) \\ P^9_{\mu\nu}(p) &= \frac{1}{24p^2}[p_\mu,\gamma_\nu] \\ P^{10}_{\mu\nu}(p) &= \frac{1}{24p^2}\left(-2\not{p}p_\mu p_\nu + p^2[p_\mu,\gamma_\nu]\right) \end{aligned} \tag{3a.7}$$

As already stated, once the the $\Delta \to 0$ limit is taken, all the above factors V, T (3a.4) depend only on the variable $u \equiv -p^2$, like the form factors

$$S_k(u),\ k = 1,2 \tag{3a.8}$$

of the self-mass amplitude:

$$i\Sigma(p) = i(i\not{p} + 1)[S_1(u) + (i\not{p} + 1)S_2(u)] \tag{3a.9}$$

where mass renormalization has been carried out, but not wave function renormalization. Of course also self-mass form factors can be obtained like (3a.4).

$$S_k(u) = Tr\ \{P^k(p)i\Sigma(p)\}, \qquad k = 1,2 \tag{3a.10}$$

and respectively:

Scalar Projectors for $i\Sigma(p)$:

$$P^1(p) = -\frac{i}{4p^2}\left((p^2-1)\frac{-i\not{p}+1}{p^2+1} + 1\right),$$
$$P^2(p) = -\frac{i}{4p^2}\left(\frac{-i\not{p}+1}{p^2+1} - 1\right) \tag{3a.11}$$

In the $\Delta \to 0$ limit the only differences between vertex and self-mass are:

i) the strings of γ-matrices in the numerator, giving different polynomials in the external and loop momenta when the form factors are projected;

ij) square denominators in the vertex amplitudes. As the main difficulties in analytical calculation come from denominators, the "core" of the analytical calculation is to handle them, and it it important to be able to handle them in a simpler form. One has, for instance, that the denominator of the electron-propagator just before and after the emission of the external photon can be expanded in Δ:

$$\frac{1}{(p \pm \frac{1}{2}\Delta)^2 + 1} = \frac{1}{p^2+1} \mp \frac{\Delta p}{(p^2+1)^2} + O(\Delta^2) \Rightarrow$$
$$\frac{1}{[(p+\frac{1}{2}\Delta)^2+1]}\frac{1}{[(p-\frac{1}{2}\Delta)^2+1]} = \frac{1}{(p^2+1)^2} + O(\Delta^2)$$

So integrals with square denominators can be calculated as derivatives of suitable single denominators

$$\int d^4p \ldots \frac{1}{(p^2+1)^2} = -\frac{\partial}{\partial \mu^2}\int d^4p \ldots \left(\frac{1}{p^2+\mu^2} + \frac{1}{p^2+M^2}\right)\Bigg|_{\mu^2=1}$$

where the exchange between integration and derivation is allowed, and Pauli-Villars counterterm is only for convergence, and does not play any other role.

When a suitable analytic expression for the self-mass form factors is obtained, vertex form factors can then be derived by "straighforward" operations, such as rearranging terms and taking derivatives with respect to intermediate masses.

The knowledge of the "off mass shell" amplitudes permits also their insertion in more complicated graphs, being careful of infrared divergencies. It becomes then possible to calculate separate single "subgraphs" and to "assemble" them in the complete graph. This technique allows, for instance, a rather simple evaluation of "multiple–ladder" graphs [9], which were the most difficult using old methods [1].

–3b) Integration

In principle one has the standard dispersion relations:

$$F(-q^2) = \frac{1}{\pi} \int \frac{du}{q^2 + u} \, \mathrm{ImF(u)} \tag{3b.1}$$

where F may be any of V, T, S (3a.4) (3a.8) and discontinuities are taken according Cutkosky–Veltman rule [11] before performing the $(-\partial/\partial\mu^2)$, if necessary. Subtracted relations are used if necessary, and, as usual, infrared singularities are parametrized giving fictious mass λ to photon.

If the number of graphs increases too much and the most complicated structure can be isolated, it may be convenient to Wick rotate dp_0 and use hyperspherical integration $\int dp_1 dp_2 dp_3 dp_4$ for peripherical propagators [8].

These techniques permit to control ellipticity and divergencies. About ellipticity, although there is a lot of radicals involving integration variables, they are all of the kind:

$$R(a, b, c) = \sqrt{a^2 + b^2 + c^2 - 2ab - 2bc - 2ac} \tag{3b.2}$$

which is the usual "phase space root". The integrations can be conveniently done by introducing the "natural variables" [1],[6],[8] which allow their analytical evaluation (if the roots are not too many!). The choice of the suitable "natural variable" depends on the integration range, determined by the values of a, b. For instance, given the complete root of eq. (3b.2), when $a \in [(\sqrt{b} + \sqrt{c})^2, \infty[$, i.e. over the threshold, one has to perform the change of variable:

$$a = +\frac{\sqrt{bc}}{x}\left(x + \sqrt{\frac{b}{c}}\right)\left(x + \sqrt{\frac{c}{b}}\right) \quad \Longleftrightarrow \quad x = \frac{+(+a - b - c) - R(+a, +b, +c)}{2\sqrt{bc}} \tag{3b.3}$$

to have:

$$R(a, b, c) = \frac{\sqrt{bc}}{x}(1 - x^2)$$

and, for instance, for a most common integral:

$$\int_{(\sqrt{b}+\sqrt{c})^2}^{+\infty} \frac{da}{R(+a, +b, +c)} f(a) = \int_0^1 \frac{dx}{x} f(a(x))$$

Such changes of variables must be done as late as possible, because due to them the number of terms increases enormously also for computer, and moreover it becomes extremely hard to control fictious divergencies arising from separate calculation of integrals. It is then essential to manipulate integrals in their elliptic aspect as far as it is possible, and to find relations to calculate analitically groups of integrals which would be one by one elliptic.

When the suitable method has been choosen, analytical calculations [1],[5],[6],[7],[8],[9] are performed using powerful and fast algebraic computer programs: SCHOONSCHIP [12], by M.Veltman, on CDC and (now) on SUN computers, and ASHMEDAI by M.J.Levine on VAX computers; numerical calculations [3] are done using the adaptative Monte Carlo integration routines RIWIAD and VEGAS [13].

−3c) Nielsen Polylogarithms

Multiple integration of logarithms required by the calculation of Feynman graphs requires the knowledge of a special class of functions, the Nielsen Polylogarithms [1],[8],[14], defined as:

$$S_{n,p}(x) = \frac{(-1)^{(n+p-1)}}{(n-1)!p!}\int_0^1 \frac{dt}{t}\,\lg^{n-1} t \lg^p(1-xt) \qquad n,p \geq 1 \tag{3c.1}$$

For $p = 1$ one defines:

$$Li_n(x) \equiv S_{n-1,1}(x) \tag{3c.2}$$

that when $n = p = 1$ reduces to the Euler dilogarithm:

$$S_{1,1}(x) = Li_2(x) = -\int_0^1 \frac{\lg(1-xt)}{t}dt = -\int_0^x \frac{\lg(1-t)}{t}dt \tag{3c.3}$$

which is stricty related to the so-called Spence funcion, defined usually as:

$$\Phi(x) \equiv \int_1^x \frac{\lg(1+t)}{t}dt = Li_2(-x) + Li_2(-1) \tag{3c.4}$$

One has the most important rules (not enough pointed out by some contributes)

$$\frac{d}{dx}S_{n,p} = \frac{1}{x}S_{n-1,p}(x) \qquad \int_0^x \frac{dt}{t}S_{n,p}(t) = S_{n+1,p}(x) \tag{3c.5}$$

The possibility to extend to usual logarithms these rules, defined above only for $n, p \geq 1$, also to $n = 0, p \geq 1$ defining

$$S_{0,p} \equiv \frac{(-1)^p}{p!}\lg^p(1-y) \tag{3c.6}$$

suggests the introduction of a classification of the "trascendentality", the Logarithmic Degree (LD):

$$LD\,\{S_{n,p}\} = n + p \tag{3c.7}$$

giving LD to products of logarithms according to the usual rule

$$LD\ \{\lg^{p_1}(\ldots)\lg^{p_2}(\ldots)\ldots\lg^{p_n}(\ldots)\} = p \quad \text{if} \quad p_1+p_2+\ldots+p_n = p$$

extended to

$$LD\ \{S_{n_1,p_1}(\ldots)\ S_{n_2,p_2}(\ldots)\} = n_1+p_1+n_2+p_2$$

and to

$$LD = 0$$

for pure algebraic expressions.
LD is conserved in algebraic transformations of the argument, like $x \to 1-x$, $x \to 1/x$, ..., integrations by parts or sums and decompositions, but, due to (3c.5), LD is increased by 1 by non trivial integrations (i.e. not by part) and is decreased by 1 by differentiation:

$$\begin{aligned} LD\left\{\int dx\Big(\frac{1}{x},\frac{1}{x-1},\frac{1}{x+1},\ldots\Big)\ S(x,LD=n)\right\} &= n+1 \\ LD\left\{\frac{d}{dx}S(x,LD=n)\right\} &= n-1 \end{aligned} \tag{3c.8}$$

This structure allows to handle "pure logarithmic expressions", i.e. expressions which are sum of addenda having the same LD (logarithmic degree). Calculation becomes then simpler and more ordered!

–3*d*) Algorithms for analytical calculation

As stated above, it is better to handle "pure logarithmic expressions" [8],[15]. After some algebraic manipulations it is possible to write the integrals in one of the "standard" forms:

$$\begin{aligned} F_1(a,y,z) &= \int_{x_1}^{x_2} \frac{dx}{R(x,y,z)} f(x,y) \\ F_2(a,y,z) &= R(a,y,z)\int_{x_1}^{x_2} \frac{dx}{R(x,y,z)}\frac{1}{(x-a)} f(x,y) \end{aligned} \tag{3d.1}$$

where R is the usual phase space root (3b.2) and $f(x,y)$ is either a rational function or

$$f(x,y) = \lg\frac{\alpha(x,y)+\beta(x,y)R(x,y,z)}{\alpha(x,y)-\beta(x,y)R(x,y,z)}$$

and α,β may be either radicals or functions having logarithmic degree lower than $F(a,y,z)$.

It is worth to differentiate them to have lower LD expressions:

$$\begin{aligned}\frac{\partial}{\partial y}F_1(a,y,z) &= \frac{\partial}{\partial y}\left\{\int_{x_1}^{x_2}\frac{dx}{R(x,y,z)}f(x,y)\right\} = \\ &\left[\left(\frac{\partial x}{\partial y}-\frac{x+y-z}{2y}\right)\right]_{x=x_1}^{x=x_2}+\int_{x_1}^{x_2}\frac{dx}{R(x,y,z)}\left\{\frac{\partial f(x,y)}{\partial y}+\frac{x+y-z}{2y}\frac{\partial f(x,y)}{\partial x}\right\}\end{aligned}$$

$$\begin{aligned}\frac{\partial}{\partial y}F_2(a,y,z) &= \frac{\partial}{\partial y}\left\{R(a,y,z)\int_{x_1}^{x_2}\frac{dx}{R(x,y,z)}\frac{1}{(x-a)}f(x,y)\right\} = \\ &\left[\left(\frac{\partial x}{\partial y}\frac{R(a,y,z)}{(x-a)}-\frac{x(-y-a+z)+3y^2-2yz-ay+az-z^2}{2yR(a,y,z)}\right)\frac{f(x,y)}{R(x,y,z)}\right]_{x=x_1}^{x=x_2} \\ &+\int_{x_1}^{x_2}\frac{dx}{R(x,y,z)}\left\{\frac{R(a,y,z)}{(x-a)}\frac{\partial f(x,y)}{\partial y}\right. \\ &\left.-\frac{x(-y-a+z)+3y^2-2yz-ay+az-z^2}{2yR(a,y,z)}\frac{\partial f(x,y)}{\partial x}\right\}\end{aligned} \tag{3d.2}$$

this must be done several times to obtain a "simple" expression, i.e. which can be integrated by means of the changes of variable of the kind of eq. (3b.3). Repeated integrations give then the final result. A fine example is [15]:

$$\begin{aligned}C = 2\int_0^\infty &\frac{dl}{\sqrt{l(l+4)}}\int_1^\infty\frac{da}{a}\frac{1}{R(a,-l,1)}\int_4^\infty\frac{db}{b}\times \\ &\left\{\frac{R(a,-l,1)-(a-l-1)}{R(-m_-,-l,b)}\lg\frac{b(b+m_-+l)+\sqrt{b(b-4)}R(-m_-,-l,b)}{b(b+m_-+l)-\sqrt{b(b-4)}R(-m_-,-l,b)}\right. \\ &\left.+\frac{R(a,-l,1)+(a-l-1)}{R(-m_+,-l,b)}\lg\frac{b(b+m_++l)+\sqrt{b(b-4)}R(-m_+,-l,b)}{b(b+m_++l)-\sqrt{b(b-4)}R(-m_+,-l,b)}\right\}\end{aligned} \tag{3d.3}$$

where R is the usual phase space root (3b.2) and

$$m_\pm = \frac{1}{2a}\{(a-1)^2+l(a+1)\pm R(a,-l,1)(a-1)\}$$

One has then to write (3d.3) as products of "standard" integrals (3d.1) multiplying and dividing by suitable square roots of the kind (3b.2) to obtain:

$$\begin{aligned}C_2(l,a) = R(0,-l,-m_-)&\int_4^\infty\frac{db}{R(b,-l,-m_-)}\frac{1}{(b-0)}\lg(\ldots\ m_-\ldots) \\ +R(0,-l,-m_+)&\int_4^\infty\frac{db}{R(b,-l,-m_+)}\frac{1}{(b-0)}\lg(\ldots\ m_+\ldots)\end{aligned} \tag{3d.4}$$

$$C_1(l) = R(1,-l,1)\int_1^\infty\frac{da}{R(a,-l,1)}\frac{1}{(a-1)}C_2(l,a)$$

$$C = R(1,-l,1)\int_0^\infty dl\left(\frac{1}{l} - \frac{1}{l+4}\right) C_1(l) \qquad (3d.5)$$

where $R(1,-l,1) = \sqrt{l(l+4)}$; $R(1,+b,1) = \sqrt{b(b-4)}$; $R(0,-l,m_\pm) = |l - m_\pm|$
Note that $LD(C_2) = 2$; $LD(C_1) = 3$; $LD(C) = 4$.
According to the rule above explained:

1) differentiate C_1 to obtain $\frac{\partial}{\partial l}C_2(l,a)$
2) differentiate C_2 to obtain "simple" integrals in b
3) integrate up to obtain again C

Despite the enormous algebra, the result is simple, as usual in QED. Defining w as "natural variable" (3b.3)

$$w = \frac{\sqrt{l(l+4)} - \sqrt{l}}{\sqrt{l(l+4)} + \sqrt{l}} \qquad l = \frac{(1-w)^2}{w} \qquad (3d.6)$$

one has

$$\begin{aligned} LD = 2 \qquad & C_2(l,a) = \frac{\lg^2 w - 2\zeta(2)}{\sqrt{l(l+4)}} \\ LD = 3 \qquad & C_1(l) = -\frac{1}{3}\lg^3 w - 2\zeta(2)\lg w \\ LD = 4 \qquad & C(l) = [3\zeta(2)]^2 = \frac{\pi^4}{4} \end{aligned} \qquad (3d.7)$$

As usual in QED calculations the result is quite simple, despite the amount of work. Is there a more compact approach to QED? The problem stands still open, also if many improvements have been made since thirty years!

Acknowledgments
The author wishes to thank Dr H. Czyż and Dr S. Laporta for having carefully read the manuscript.

References

[1] R.Barbieri, J.A.Mignaco and E.Remiddi: Nuovo Cim. 11A (1972) 824.
[2] R.S.Van Dyck, Jr., P.B.Schwinger H.G.Dehmelt: Phys. Rev. Lett 59 (1987) 26.
[3] T.Kinoshita: in "The Hydrogen Atom", Springer Verlag, Berlin (1989) 247.
[4] J.Schwinger: Phys. Rev. 73 (1948) 416, 76 (1949) 790;
R.Karplus and N.M.Kroll: Phys. Rev. 77 (1950) 536;
A.Petermann: Helv. Phys. Acta 30 (1957) 407;
C.M.Sommerfield: Ann of Phys. 5 (1958) 26.
[5] R.Barbieri and E.Remiddi: Nucl. Phys. B 90 (1975) 233
[6] R.Barbieri, M.Caffo, E.Remiddi, S.Turrini and D.Oury:
Nucl. Phys. B144 (1978) 329.
[7] M.J.Levine, R.Roskies, Phys. Rev. D 9 (1974) 421;
M.J.Levine, R.C.Perisho, R.Roskies, Phys. Rev. D 13 (1976) 997;
M.J.Levine, R.Roskies, Phys. Rev. D 14 (1976) 2191.
[8] M.Caffo, E.Remiddi and S.Turrini: Nuovo Cim. 79A (1984) 220.
[9] M.Caffo, E.Remiddi and S.Turrini: Nucl. Phys. B141 (1978) 302;
M.Caffo, S.Turrini and E.Remiddi: Phys. Rev. D 30 (1984) 483;
E.Remiddi and S.Sorella, Lett. Nuovo Cim. 44 (1985) 231.
[10] M.E.Cage *et al.*, presented at the 1988 Conference on Precision Electromagnetic Measurements, Tsukuba, Japan, June 7–10, 1988.
[11] R.E.Cutkosky: Journ. of Math. Phys. 1 (1960) 429
M.Veltman "Diagrammar" C.E.R.N. Th.D. 73-9 (1973)
[12] H.Strubbe: Comp. Phys. Comm. 8 (1981) 1.
[13] B.E.Lautrup: CERN Computer Center Program Library, subroutine D114 (1983);
G.P.Lepage: J. Comp. Phys. 27 (1978) 192.
[14] N.Nielsen: Nova Acta 90 (Halle 1909) 125;
K.S.Kölbig, J.A.Mignaco and E.Remiddi: BIT 10 (1970) 38.
[15] M.Caffo, E.Remiddi and S.Turrini: in Lecture Notes in Computer Science 72, Springer Verlag, Berlin (1979) 52.

LIST OF PARTICIPANTS

A. Amoroso
Department of Mathematics
University of Bologna
piazza Porta San Donato 5
40126 Bologna (Italy).

B. Arcipiani
ENEA
Centro Ricerche Energia "Casaccia"
via Anguillara 301
00060 S.Maria di Galeria, Rome (Italy).

F.T. Arecchi
Department of Physics
University of Florence, and
Istituto Nazionale di Ottica
largo E.Fermi 6
50125 Florence (Italy).

E. Baicchi
ENEA
Centro Ricerche Energia "Brasimone"
40032 Camugnano, Bologna (Italy).

L. Baldi
ENEA
Centro Ricerche Energia "Brasimone"
40032 Camugnano, Bologna (Italy).

L. Battiston
C.N.R. - I.S.D.G.M.
S.Polo 1364
30125 Venice (Italy).

M. Benzi
Department of Mathematics
University of Bologna
piazza Porta San Donato 5
40126 Bologna (Italy).

V. Benzi
Istituto di Ingegneria Nucleare
Facolta' di Ingegneria
University of Bologna
40136 Bologna (Italy).

M. Biasini
ENEA
Centro Ricerche Energia "E.Clementel"
viale Ercolani 8
40138 Bologna (Italy).

R. Bonetti
Istituto di Fisica Generale
University of Milan
via Celoria 16
20133 Milan (Italy).

G. Bonsignori
Department of Physics
University of Bologna
via Irnerio 46
40126 Bologna (Italy).

E. Bortolani
ENEA
Centro Ricerche Energia "E.Clementel"
viale Ercolani 8
40138 Bologna (Italy).

R. Borsari
Istituto di Ingegneria Nucleare
Facolta' di Ingegneria
University of Bologna
40136 Bologna (Italy).

A. Brini
Department of Mathematics
University of Bari
Bari (Italy).

U. Broccoli
ENEA
Centro Ricerche Energia "E.Clementel"
via Martiri di Monte Sole
40138 Bologna (Italy).

A. Cardinali
ENEA
Centro Ricerche Energia "Frascati"
via E.Fermi 27
00044 Frascati, Rome (Italy).

G. Cenacchi
ENEA
Centro Ricerche Energia "E.Clementel"
viale Ercolani 8
40138 Bologna (Italy).

Cernuschi
Department of Physics
University of Milan
via Celoria 16
20133 Milan (Italy).

A. Cherubini
ENEA
Centro Ricerche Energia "E.Clementel"
viale Ercolani 8
40138 Bologna (Italy).

C. Chiesa
Department of Physics
University of Milan
via Celoria 16
20133 Milan (Italy).

C. Chiccoli
CNAF and Istituto Nazionale di
Fisica Nucleare, Sezione di Bologna
viale Ercolani 8
40138 Bologna (Italy).

F. Cleri
ENEA
Centro Ricerche Energia "Casaccia"
via Anguillara 301
00060 S.Maria di Galeria, Rome (Italy).

E. Cupini
ENEA
Centro Ricerche Energia "E.Clementel"
viale Ercolani 8
40138 Bologna (Italy).

V. Cuzzola
ENEA
Centro Ricerche Energia "Brasimone"
40032 Camugnano, Bologna (Italy).

G. Dattoli
ENEA
Centro Ricerche Energia "Frascati"
via E.Fermi 27
00044 Frascati, Rome (Italy).

D'Arrigo
Department of Physics
University of Messina
via Tommaso Cannizzaro
98100 Messina (Italy).

G. De Canio
ENEA
Centro Ricerche Energia "Casaccia"
via Anguillara 301
00060 S.Maria di Galeria, Rome (Italy).

P. De Chiara
Department of Physics
University of Ferrara
via Paradiso 12
44100 Ferrara (Italy).

A. De Matteis
ENEA
Centro Ricerche Energia "E.Clementel"
viale Ercolani 8
40138 Bologna (Italy).

S. De Nicola
C.N.R. - Istituto di Cibernetica
via Toiano 2
80072 Arco Felice, Naples (Italy).

M. Errani
CNAF and Istituto Nazionale di
Fisica Nucleare, Sezione di Bologna
viale Ercolani 8
40138 Bologna (Italy).

F. Fabbri
ENEA
Centro Ricerche Energia "E.Clementel"
viale Ercolani 8
40138 Bologna (Italy).

J. Fernandez
Istituto di Ingegneria Nucleare
Facolta' di Ingegneria
University of Bologna
40136 Bologna (Italy).

R. Forghieri
ENEA
Centro Ricerche Energia "E.Clementel"
viale Ercolani 8
40138 Bologna (Italy).

G. Forni
Department of Mathematics
University of Bologna
piazza Porta San Donato 5
40126 Bologna (Italy).

M. Frisoni
ENEA
Centro Ricerche Energia "E.Clementel"
viale Ercolani 8
40138 Bologna (Italy).

L. Fronzoni
Department of Physics
University of Calabria
87036 Arcavacata di Rende, Cosenza
(Italy).

L. Galgani
Department of Mathematics
University of Milan
via Saldini 50
20133 Milan (Italy).

M. Galli
ENEA
Centro Ricerche Energia "E.Clementel"
viale Ercolani 8
40138 Bologna (Italy).

M. Giovannozzi
Department of Physics
University of Bologna
via Irnerio 46
40126 Bologna (Italy).

D. Giusti
Istituto di Ingegneria Nucleare
Facolta' di Ingegneria
University of Bologna
40136 Bologna (Italy).

Guidi
Department of Physics
University of Ferrara
via Paradiso 12
44100 Ferrara (Italy).

M. Herman
ENEA
Centro Ricerche Energia "E.Clementel"
viale Ercolani 8
40138 Bologna (Italy).

F. Iachello
Sloane Center for Theoretical Physics
Yale University
New Haven CT (U.S.A.).

R. Dos Anjos Lima
Centre de Physique Theorique
C.N.R.S. Luminy Case 907
13288 Marseille (France).

N. Lo Iudice
Department of Physics
University of Naples
Mostra d'Oltremare - pad.20
80125 Naples (Italy).

C. Lolli
ENEA
Centro Ricerche Energia "E.Clementel"
viale Ercolani 8
40138 Bologna (Italy).

L. Lorenzelli
ENEA
Centro Ricerche Energia "Brasimone"
40032 Camugnano, Bologna (Italy).

Lorenzoni
Department of Physics
University of Ferrara
via Paradiso 12
44100 Ferrara (Italy).

S. Lorenzutta
ENEA
Centro Ricerche Energia "E.Clementel"
viale Ercolani 8
40138 Bologna (Italy).

V. Loschiavo
C.N.R. - Istituto di Fisica
dello Spazio Interplanetario
Casella Postale 27
00444 Frascati, Rome (Italy).

G. Lo Vecchio
Department of Physics
University of Ferrara
via Paradiso 12
44100 Ferrara (Italy).

A. Luise
ENEA
Viale Regina Margherita
00100 Roma (Italy).

F. Mainardi
Department of Physics
University of Bologna
via Irnerio 46
40126 Bologna (Italy).

G. Maino
ENEA
Centro Ricerche Energia "E.Clementel"
viale Ercolani 8
40138 Bologna (Italy).

V.R. Manfredi
Department of Physics
University of Padua
via Marzolo 8
35131 Padua (Italy).

M. Marangoni
ENEA
Centro Ricerche Energia "E.Clementel"
viale Ercolani 8
40138 Bologna (Italy).

L. Martelli
ENEA
Centro Ricerche Energia "E.Clementel"
viale Ercolani 8
40138 Bologna (Italy).

E. Menapace
ENEA
Centro Ricerche Energia "E.Clementel"
viale Ercolani 8
40138 Bologna (Italy).

A. Mengoni
ENEA
Centro Ricerche Energia "E.Clementel"
viale Ercolani 8
40138 Bologna (Italy).

V. Molinari
Istituto di Ingegneria Nucleare
Facolta' di Ingegneria
University of Bologna
40136 Bologna (Italy).

C. Nappi
C.N.R. - Istituto di Cibernetica
via Toiano 2
80072 Arco Felice, Naples (Italy).

M. Ottaviani
ENEA
Centro Ricerche Energia "Frascati"
via E.Fermi 27
00044 Frascati, Rome (Italy).

P.L. Ottaviani
ENEA
Centro Ricerche Energia "E.Clementel"
viale Ercolani 8
40138 Bologna (Italy).

Paoli
Istituto Nazionale di Ottica
largo E.Fermi 6
50125 Florence (Italy).

P. Pasini
CNAF and Istituto Nazionale di
Fisica Nucleare, Sezione di Bologna
viale Ercolani 8
40138 Bologna (Italy).

M. Pettini
Osservatorio Astrofisico di Arcetri
largo E.Fermi 5
50125 Florence (Italy).

F. Pizzio
ENEA
Centro Ricerche Energia "Casaccia"
via Anguillara 301
00060 S.Maria di Galeria, Rome (Italy).

A. Politi
Istituto Nazionale di Ottica
largo E.Fermi 6
50125 Florence (Italy).

G. Pozzi
Department of Physics
University of Lecce
via Arnesano
73100 Lecce (Italy).

F. Premuda
Istituto di Ingegneria Nucleare
Facolta' di Ingegneria
University of Bologna
40136 Bologna (Italy).

I. Purica
ENEA
Viale Regina Margherita
Rome (Italy).

A. Quintiliani
ENEA
Viale Regina Margherita
Rome (Italy).

M. Rasetti
Department of Physics
Politecnico di Torino
corso Duca degli Abruzzi 24
10129 Torino (Italy).

G. Reffo
ENEA
Centro Ricerche Energia "E.Clementel"
viale Ercolani 8
40138 Bologna (Italy).

E. Remiddi
Department of Physics
University of Bologna
via Irnerio 46
40126 Bologna (Italy).

F. Romanelli
ENEA
Centro Ricerche Energia "Frascati"
via E.Fermi 27
00044 Frascati, Rome (Italy).

E. Santamato
Department of Physics
University of Naples
Mostra d'Oltremare, pad. 20
80125 Naples (Italy).

E. Santoro
ENEA
Centro Ricerche Energia "Casaccia"
via Anguillara 301
00060 S.Maria di Galeria, Rome (Italy).

A. Sargeni
ENEA
Centro Ricerche Energia "Casaccia"
via Anguillara 301
00060 S.Maria di Galeria, Rome (Italy).

F. Semeria
CNAF and Istituto Nazionale di
Fisica Nucleare, Sezione di Bologna
viale Ercolani 8
40138 Bologna (Italy).

M. Stefanon
ENEA
Centro Ricerche Energia "E.Clementel"
viale Ercolani 8
40138 Bologna (Italy).

G. Stoppini
Department of Physics
University of Pisa
piazza Torricelli
56100 Pisa (Italy).

M. Sumini
Istituto di Ingegneria Nucleare
Facolta' di Ingegneria
University of Bologna
viale Risorgimento 2
40136 Bologna (Italy).

A. Taccone
Department of Physics
University of Messina
via Tommaso Cannizzaro
98100 Messina (Italy).

C. Tebaldi
Facolta' di Ingegneria
University of Bologna
via Saragozza 8
40100 Bologna (Italy).

E. Todisco
Department of Physics
University of Bologna
via Irnerio 46
40126 Bologna (Italy).

L. Tondinelli
ENEA
Centro Ricerche Energia "Casaccia"
via Anguillara 301
00060 S.Maria di Galeria, Rome (Italy).

A. Torre
ENEA
Centro Ricerche Energia "Frascati"
via E.Fermi 27
00044 Frascati, Rome (Italy).

T. Trombetti
Istituto di Ingegneria Nucleare
Facolta' di Ingegneria
University of Bologna
40136 Bologna (Italy).

G. Turchetti
Department of Physics
University of Bologna
via Irnerio 46
40126 Bologna (Italy).

S. Turrini
Department of Physics
University of Bologna
via Irnerio 46
40126 Bologna (Italy).

M. Vaccari
ENEA
Centro Ricerche Energia "E.Clementel"
viale Ercolani 8
40138 Bologna (Italy).

M. Valli
ENEA
Centro Ricerche Energia "E.Clementel"
viale Ercolani 8
40138 Bologna (Italy).

A. Ventura
ENEA
Centro Ricerche Energia "E.Clementel"
viale Ercolani 8
40138 Bologna (Italy).

G. Vitali
Istituto di Agronomia
University of Bologna
via Filippo Re 6/8
40100 Bologna (Italy).

T. Volpe
ENEA
viale Regina Margherita
00100 Roma (Italy).

S. Zerbini
Department of Physics
University of Trento
38050 Povo, Trento (Italy).

L. Zuffi
Department of Physics
University of Milan
via Celoria 16
20133 Milano (Italy).

AUTHOR INDEX

www.ingramcontent.com/pod-product-compliance
Lightning Source LLC
LaVergne TN
LVHW071223011026
849280LV00007B/40